Method for engineering students

Degree Projects Using the 4-phase Model

PÄR BLOMKVIST
ANETTE HALLIN

Originally published by Studentlitteratur AB in Swedish under the title *Metod för teknologer.*

Copying prohibited

All rights reserved. No part of this publication may be reproduced or transmitted in any form or by any means, electronic or mechanical, including photocopying, recording, or any information storage and retrieval system, without permission in writing from the publisher.

The papers and inks used in this product are eco-friendly.

Art. No 38244
ISBN 978-91-44-09555-4
First edition
1:1

© The authors and Studentlitteratur 2015
www.studentlitteratur.se
Studentlitteratur AB, Lund

Cover design: Carina Blomdell
Cover photo: Martine Castoriano

Printed by Graficas Cems S.L., Spain 2015

CONTENTS

Preface 5

Introduction 7
    Engineering students and the social sciences 7
    The degree project – craftsmanship 10

Phase 1 Formulate 19
Prototype 1:1 Problematization (not just problem-solving!) 21
    Working documents 24
    Purpose and question formulation 26
    Applying a system perspective 29
    Divergent and convergent thinking 31
    Independence and managing (dual) clients 33
    Tools for thought – various knowledge interests 36
    Ethics 37
    Learning goals as a checklist 39
Prototype 1:2 Research proposal – explicitly defining and making your problematization scientific 39
    Searching and relating to literature 43
    What is theory? 46
    Relationship between theory and empirics: induction, deduction, abduction 48
    Scientific writing 48
    What is science and how do we measure scientific quality? 52
    Critical thinking – the basis of science 54

Phase 2 Design 57
Prototype 2:1 Mid-term report 59
    On qualitative and quantitative method 61
    Choice of research design 62
    Choice of data gathering method 74

Phase 3 Produce 97

Arrangement and content of the academic paper 98
Prototype 3:1–3:5 (× n) 101
    Prototype 3:1: write an expanded synopsis 103
    Prototype 3:2: write a version of the second part of the thesis 104
    Prototype 3:3: write a version of the third part of the thesis 105
    Prototype 3:4: write a new version of the first part of the thesis 106
    Prototype 3:5: work through the language, layout, and formalities 107
    More iterations – more prototype attempts 108
    On the analysis of empirics 114
    On sources and reference management 122
    Argumentation and a thesis driven paper 129
    The degree project in a wider context – science and society 130

Phase 4 Deliver 137

Prototype 4:1–4:2 138
    Prototype 4:1: oral presentation before the thesis is fully finished 139
    Prototype 4:2: presenting at a final seminar 140
    Presentations of the finished product 142
    Template for an argumentative talk 143
    Template for analysing academic texts 146

Conclusion 149

Further reading 153

Subject index 157

PREFACE

This book presents a model for degree projects based on two key principles: *prototyping* and *presenting*. This is also how we have worked as authors. The text has gone back and forth between us many times, with the prototypes numbering at least 50 or so at this stage. We have worked together and fully assume joint responsibility for the entire text, even though Pär Blomkvist is the creator of the 4-phase model. We have also had the privilege of presenting the text in various contexts, receiving valuable input which we have benefitted from when writing. In January 2014, we presented a relatively complete prototype to the IDA seminar at the Department of Industrial Economics and Management, Royal Institute of Technology, Stockholm, obtaining important help from our brilliant opponents Håkan Kullvén and Pernilla Ulfvengren.

Our thanks go to our colleagues at INDEK, who have read and commented on the entire manuscript: Niklas Arvidsson, Cali Nuur, Staffan Laestadius, Mats Engwall, Johann Packendorff, Charlotta Linse, Fabian Levihn and Stefan Tongur. We would also like to thank Kent Thorén and Mandar Dabhilkar, who have provided us with valuable feedback on the sections on quantitative method, as well as Tina Karrbom-Gustavsson and Åke Walldius, who read the introductory sections particularly carefully.

Out thanks also go to Susanne Blomkvist, Johan Flygare and Anneli Flygare Rydell, who with their customary acuity read the entire text.

Our special thanks go to Lars Uppvall who, together with Pär Blomkvist, theoretically as well as practically during course development work, was involved in creating several important components of the 4-phase model. Lars has also critically reviewed the manuscript and has also used a prototype of the book when supervising degree projects.

Finally, we would like to extend our warm gratitude both to our publisher Ola Håkansson and to our editor Åsa Sterner.

Stockholm, May 2014
Pär Blomkvist and *Anette Hallin*

INTRODUCTION

This book is primarily aimed at those studying at an institute of technology who will be writing an academic paper, often called a thesis, leading to a degree on the Bachelor's or Master's levels. We account for a process model which is general to all types of thesis work; however, the examples and methods we present here mostly originate from the field of social sciences. The book has four starting points:

- the need to know about the social sciences' research tradition
- the fact that engineering students often have two clients
- the fact that the degree project process is not linear
- the fact that the degree project is a piece of craftsmanship.

ENGINEERING STUDENTS AND THE SOCIAL SCIENCES

Firstly, we can point out that, for a long time, engineering students have often written their degree projects within or close to the field of social sciences. It is not unreasonable to assert that many a modern engineer is actually a kind of social engineer working interdisciplinarily in a field where natural science and engineering science meet society, or that there is a need for increased knowledge in the research tradition of the social sciences. This book is the first coherent set of instructions regarding how to design such a degree project. We start out from a wide definition of the social sciences, in which we include, for instance, management and organization theory, innovation theory, and engineering and industrial history. This book is envisaged for students of various disciplines within engineering, but we are convinced that it will also suit, for instance, degree courses in business administration at both universities and business schools.

DOUBLE CLIENTS

The second starting point of this book is the fact that many students' writing degree projects have *double clients*. On the one hand, they have a supervisor at university who wants a paper meeting requirements regarding scientific legitimacy, and which can thus be assessed in accordance with the grading criteria and learning goals that exist for degree projects. On the other hand, they often have a client at a company or organization who has formulated an assignment that the student is to carry out within the framework of the degree project, an assignment that entails the student him-/herself having to formulate the problem and subsequently look for a solution. Frequently, the student feels that academia and industry have differing demands regarding the content of the thesis. A common perception is that the university wants abstract theory while the company wants applicable results. We realise that this disparity can exist but we claim that it is considerably less than it may seem.

> Many engineers work in a field where natural science and engineering science encounter society.

Regardless of whether you are writing your degree project on the basis of an assignment given to you by a company, or some other organization outside the university, if your degree project is based on a supervisor's idea or originates from your own ideas, the purpose of the academic paper is to safeguard the quality of your findings. The findings that you as a degree project author deliver must be well-grounded in theory, previous research, and your own empirics (i.e. the empirical material and data you obtain via the study you are conducting). Moreover, your approach and methods must be reported in such a way that your clients – and here we mean both clients – will be able to determine whether or not your findings are reasonable. All your credibility as a researcher (or consultant or investigator, for that matter) is riding on the academic paper, even though not everyone will be reading it in full. The academic paper is the mother of all documents and something you can show if your findings are ever called into question.

THE NON-LINEAR PROCESS AND THE PROTOTYPING

The third starting point is that many students relate to working on their thesis as though it were a linear process. However, writing degree projects is neither a linear nor a sequential process. You cannot finish writing the various parts of your paper and then put them together like Lego bricks. Instead, you have to constantly link back to what you did previously, making sure that there is conformity between the various parts of your work while your paper is emerging.

We use the term *prototyping* in order to emphasize that you, as the author, are gradually building a product by working up different versions of the text, but with all parts having to be cohesive – and with you having to bear in mind the finished product, i.e. your academic paper/thesis, the whole time. You start prototyping your paper right from day one, and it has to bring your client some form of benefit. All its parts have to be cohesive, and everything that is not necessary as regards supporting your findings has to be removed from the final version. Prototyping includes giving regular presentations to other stakeholders, i.e. supervisors, external clients, and fellow students. By means of repeated presentations, you will reconcile the direction of your work and you will be able to advance your argumentation.

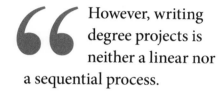

However, writing degree projects is neither a linear nor a sequential process.

In order to help you to do this, we present a model, the *4-phase model*, which is based on the notion of degree projects as a product design process whereby the goal is to develop, through gradual prototyping and presenting, a product: i.e. your academic paper/thesis. In the 4-phase model, the degree project is permeated by two key, recurring activities: prototyping and presenting. Focusing on prototyping and presenting has two advantages over a traditional approach to scientific writing and research.

Firstly, prototyping entails a much more public and social way of working. You will frequently (as frequently as possible) be showing your prototypes to and discussing them with supervisors, external clients, and fellow students. In our experience, prototyping creates a more relaxed attitude towards the different texts constituting the prototypes at their different stages of perfection. Quite simply, it is easier to take and give criticism concerning a sturdy prototype compared to a traditional section of text that you have authored yourself in solitude. Prototyping and presenting are aimed at playing down the writing of texts and showing them to others. We want an end to the image of the solitary author; instead, we borrow metaphors and analogies from creative and collective work processes such as product design, complex projects, and (why not) theatre and music. In all these fields, sketches, prototypes, and incomplete versions are shown to all those involved. No version is particularly good initially, but the aim is to assist in its improvement via criticism, prototyping, and feedback.

Secondly, the prototype process also means, right through your degree project, that you will be working in a focused way towards your final result: i.e. the product. This focus makes it easier for you, as an author, to delete, modify, and change your texts. It is not the individual section of text that counts, but the whole. In the finished product, all parts must be coherent and subordinate to the purpose of the thesis. For example, it is not important how good a certain chapter actually is. If that chapter is too long, too imprecise, and too unfocused, it must be rewritten. Designing a thesis means moulding the different parts together into a smoothly-functioning unit; then, the parts must be coherent, subordinate to the whole and work together.

> In the 4-phase model, the degree project is permeated by two key, recurring activities: prototyping and presenting.

THE DEGREE PROJECT – CRAFTSMANSHIP

This leads us to the fourth and final starting point of the book: the fact that prototyping and working on the *degree project constitutes craftsmanship* in which you as a student have to learn to understand and apply certain tactics, methods, and concepts in order to achieve a good result. Your craftsmanship and the product constituting the result – the academic paper – have their special, rather strict and at times possibly boring frameworks and rules. The standardised format aims at facilitating comparison between different products and safeguarding the quality of the result. The framework and rules also enable an opponent to critically review the arguments and reasoning, something that also quality assures the results.

HAVING A CLIENT

The 4-phase model is well-suited to a situation where you have been commissioned by a company or some other external client. But it is suitable even when you do not have an actual client as such. We claim that the degree project author should always work as though he/she had an external client. If you do not have a real client, we feel that you should be commissioned by a fictitious one. In formulating a problem for a client, the paper acquires a clear aim and focus. You obtain a reader. Furthermore, you can always send your degree project to your fictitious (but existing) client when looking for a job.

Note that the client, paradoxically enough, often has an unclear problem formulation while at the same time being strongly solution-oriented. The client does not really know what the problem is but often has thoughts about where the solution may be looked for. In such cases, it will be a matter of you, as a degree project author, taking a step backwards and really thinking about your own problematization. This should preferably provide the client with more than he/she was expecting, in addition to meeting academic demands. We are of the opinion that the same phenomenon can also arise when your problematization emanates from research or comes from your own ideas. It is very tempting to be solution-oriented all too early on. Having a client (a clear reader) also means that your degree project must be argumentative.

> Having a client also means that your degree project must be argumentative.

We feel that you, as a researcher, are obliged to argue, using scientifically well-grounded facts, that your academic paper, findings, and proposals are reasonable. To this, we also add a demand for honesty, openness, and transparency. A reader or opponent must be able, without effort, to identify delimitations and shortcomings, as well as its merits. Exactly how you write a thesis that meets these demands is what this book is all about.

YOUR TASK

The task of a degree project could be expressed in the following condensed way:

1. Formulate a non-trivial and researchable problem, sometimes as commissioned by a client who often has a diffuse set of problems and who is also simultaneously strongly solution-oriented.
2. Present a credible and well-grounded solution proposal based on a scientific investigation.

Using the 4-phase model, you obtain a structure for your work process and an overview of the fields within scientific method that you need be acquainted with in order to be able to deliver your degree project.

THIS BOOK'S APPROACH

In order to demonstrate how you as the author of the thesis have to go about designing your degree project, the approach of this book adheres to the four phases of the 4-phase model:

Phase 1: Formulate
Phase 2: Design
Phase 3: Produce
Phase 4: Deliver

When prototyping and presenting your thesis during these phases, you will need to become absorbed in various questions concerning scientific legitimacy, method issues, and the use of literature and sources. We regard these as the *scientific tools* that you need in order to design your paper.

> **The scientific tools**
>
> The scientific tools are the concepts, methods, and ways of thinking that you will need when writing a scientific paper. They are presented in connection with the prototype during which we believe that you might have the greatest need to learn about them. We run through all the tools (see the list of contents), but you may also need to consult other theory and method literature. Suggestions for this kind of literature may be found at the end of the book in the section "Further reading".

In connection with the various phases, we take up the scientific tools in the order that we assess them as being needed. The tools must support the prototyping. The tools we introduce during Phase 1 mainly focus on the introduction of the work process and contain practical advice and information. During Phases 2–4, the scientific tools gradually become adapted to the finished product and are aimed at safeguarding the quality of the thesis.

However, as with all divisions, the logic we have chosen solves certain problems while creating others. This means that you, the reader, will sometimes have to jump between different sections of the book. For instance, we have linked some important scientific tools and the templates for an academic paper with Phase 3, but we still recommend that you read these sections earlier. Maybe this will not be the first thing you do during your degree project, but somewhere around the middle of the process, you should get involved in how your final result will look and how it will be quality assured.

We do not intend to provide an exhaustive description of everything that can be said about scientific legitimacy, literature, method, and empirics. This would be impossible since each section in itself would constitute a book in its own right. Instead, we would like you, as the author, to understand how the tools function in your research process and thesis-writing. Once you have chosen your own theory and method etc, there will be other more specialized and penetrating publications that you may need to use. Some examples of further reading are thus available at the end of the book. We propose that you and your supervisor discuss which scientific tools are of key importance to you and that you read up on these in more detail.

We would like you to understand the functions of the tools.

We will constantly be providing concrete examples from previous degree projects. These examples are authentic, but anonymous, and build both on our own and on our colleagues' experiences when supervising degree projects.

You will notice that the book, in connection with the process description, contains a number of templates and proposals regarding workflows and seminars. We know that different seats of learning have different templates, processes, and series of seminars for thesis work and that these can sometimes differ depending on whether it is a Bachelor's or a Master's thesis. But we believe that the 4-phase model can still be used in its fundamentals and modified in its details in order to fit in everywhere.

THE THESIS PROCESS ACCORDING TO THE 4-PHASE MODEL

The backbone of the 4-phase model is constituted, as previously mentioned, by the four phases, i.e. Formulate – Design – Produce – Deliver. These will be dealt with in that order:

- During Phase 1, *Formulate*, you will be working up two prototypes – Prototypes 1:1–1:2 (the figure is approximate – the number of prototypes can vary between seats of learning). You will discuss these prototypes with your supervisor and probably present them at a seminar.
- During Phase 2, *Design*, you gather your materials, read up thoroughly on literature, theory, and method and use your previous prototypes to design your mid-term report: Prototype 2:1.
- During Phase 3, *Produce*, it is a matter of creating, from a number of gathered materials and ideas that have been generated, the finished product on the basis of the academic paper template applicable at your university. Phase 3, too, includes a series of different prototypes, prototype attempts, and presentations of different versions of the thesis – Prototype 3:1–3:5 (here, too, the number of prototypes can vary).
- During Phase 4, *Deliver*, you polish your paper and prepare different types of presentations of your work – both written and oral – in front of different audiences. Here, too, we regard different versions of the paper and different forms of presentations as continued prototyping – Prototype 4:1–4:2 (or more) – up until the product is completed.

THE PROCESS–PRODUCT PARADOX

As a reader of this book, you will surely understand that we, as the authors, have ended up in a paradox: How can we argue in favour of the design process being iterative and providing feedback while presenting a sequential model in which we claim that the four phases have to be carried out in a certain order? The answer is, and this is something you will also discover when working with your thesis, that the written text and the genre of the academic paper will force you into a sequential way of narrating and arguing. You have to choose a certain order so as to clarify a course of events even though that order, to a certain degree, contradicts your message. In an academic paper, this phenomenon is extra clear since that kind of product has to adhere to a strict and standardized arrangement.

CONTINUOUS FEEDBACK

In order to get us out of this paradox, to some degree, we want to emphasize that, even if the phases are worked through in order, the work process shown in Figure 1 will not be entirely modularized. In practice, the phases overlap each other. It is not possible to completely finish one phase and to then begin on another. Continuous feedback-loops between the four phases are necessary in order to achieve a good thesis. In connection with each feedback-loop, you re-write your text; in this manner, you refine the prototype on the basis of the knowledge and insight you have acquired since the last version. It is a good idea to read through the entire book to gain an understanding of the process – the entire 4-phase model – before starting on your work. In order for your thesis to turn out well, the scientific tools should not either be treated as demarcated building blocks in a linear chain. Of course, at some point, you will have to call a halt and deliver the thesis with the theory, literature, methods, and empirics that you have. But that point must not come too early. By continuously returning to the tools along your way, you will enhance your knowledge of scientific tools, improve your text, and thus strengthen your argumentation. That is to say: continuous feedback and feedback-loops also apply both to the tools and to their relationship with the four phases of the model.

In brief, you must, for as long as possible, *retain the uncertainty or ambiguity* – change your mind – and allow yourself to be surprised by what you read or run into during your investigation. You are doing feedback-based work in which the reflections you have when switching backwards and forwards between what you are currently doing, the feedback you get from others about your prototypes, and the knowledge you are developing about the tools you are using will lead to increasingly better versions of your project work.

> Continuous feedback-loops between the four phases are necessary in order to achieve a good thesis.

> By continuously returning to the tools along your way, you will enhance your knowledge of scientific tools, improve your text, and thus strengthen your argumentation.

PRESENTATIONS AND FEEDBACK

The feedback-loops are not just created between your own feedback between previous prototypes of the text and the prototype you are working with – or between the new insights you gain when your knowledge of the scientific tools is enhanced and previous thoughts – but also between the reflections and input you get from others who read and comment on the prototype you are presenting, e.g. at a seminar. The fact is that one key dimension of the 4-phase model is the creation of a context whereby collective efforts and collaborations with other students, as well as with supervisors and other stakeholders, effectively help you to succeed in writing a thesis that is as good as possible. By presenting the prototypes at different seminars where you get feedback (and provide it to your fellow students), you will obtain valuable knowledge that you can subsequently use to write new material and improve your prototype. One often-forgotten aspect is time. You have to give feedback, go deeper into your thoughts, and be open to the comments and input that you receive from others. However, you also have to give yourself time to process new impressions. Sleeping on it, giving yourself some breathing space, and doing something else all constitute, at the very minimum, an equally important part of the work process. Our advice is to consciously and clearly plan such pauses for reflection into your schedule. In addition to your presentation of your prototypes at the seminars set by your university, we would strongly recommend you, as the author, to jointly form *ad hoc* reading groups on the side with your fellow students. In industry and academia, such external groups are common in projects. These can, for instance, be called *reference groups* or advisory boards. By regularly presenting your prototypes to others, you will develop your thinking and refine your findings. Back at your place, and in your own mind, not enough is going on. The real boost occurs when you have an audience who see what you have prototyped and who give you feedback so you can improve and refine that prototype. One general tip is to write a *research diary* or log right from the start of your degree project. Here, you jot down ideas, suggestions, critical viewpoints, notes from when you change the focus of your investigation, and so on. You will benefit greatly from this diary at the end of the process as it is very easy to forget what you have done. We recommend that you also, in consultation with your supervisor, contact *specialists* in your field during the course of the process. These can factually scrutinize your arguments, evidence, and findings. You can seek specialist competence both at the university and within trade and industry. Trade associations and authorities are other sources of specialist knowledge. You will find that many people are happy to spare their time when it comes to degree projects.

> By regularly presenting your prototypes to others, you will develop your thinking and refine your findings.

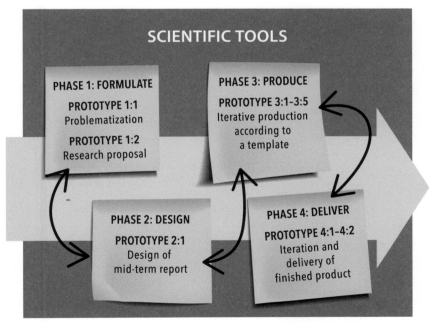

FIGURE 1 The backbone and the scientific tools of the 4-phase model. The basic idea of the model is continuous feedback between Phases 1–4 and between the respective phase and the scientific tools. The green arrow depicts time. The various phases correspond approximately to the following percentages of the degree project process over time: Phase 1: 20 %, Phase 2: 30 %, Phase 3: 35 %, Phase 4: 15 %. Note that the boundaries between the phases are not entirely distinct. The phases have been named after the activity that is dominant during the course of the process.

Phase 1 Formulate

The first phase of the thesis process is *Formulate*. This means gradually working up various prototypes focusing on formulating a scientific and researchable problem. The prototypes will then form the basis of the design work (Phase 2), of the production (Phase 3), and of the delivery and the presentations you make orally and in writing (Phase 4). During the formulation phase, you will constantly need to develop your insight into, as well as your knowledge of, what scientific work entails, how you are to relate to the literature, and what you have to do to carry out the study. In order to do this, you will need to get involved, in parallel with working on the various prototypes, in the different tools shown in Figure 2.

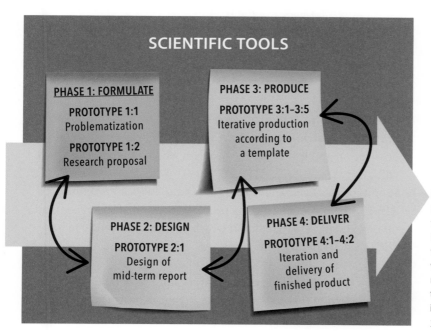

FIGURE 2 **Phase 1: Formulate**. The scientific tools are described in the book in connection with the prototype in which we believe the maximum need exists to learn about them. Phase 1 is explained in more detail in Figures 3 and 7.

The most important thing during the Formulate Phase is to make sure, for as long as possible, that you have an open attitude towards your work, while simultaneously clearly expressing, in the various prototypes, what you want to do. This may sound like a contradiction but is really a matter of, on the one hand, being open to the possibility that your purpose, question formulations, empirics, interpretations and so on may change several times during the course of prototyping, and, on the other hand, writing up, during the various prototypes, increasingly precise formulations in order to test them, both as regards yourself and vis-à-vis your supervisor, external client, and fellow students. This is called *preserving ambiguity*. If you finalize your purpose etc. too early during the process, you will surely arrive at a suboptimum result. Your investigation will become trivial if you do not allow yourself to be surprised by what you discover along the way. This applies to what you read, unexpected data and facts from interviews, other sources as well as comments from seminars and presentations. At the same time, however, you need to specify your purpose, question formulations, and method etc in order to test them.

> Keep an open attitude towards your work, but express clearly, in the various prototypes, what you want to do.

As shown in Figure 2, the Formulate Phase consists of work on two separate prototypes, i.e. gradually more and more completed versions of your product (the thesis). It will be appropriate if you write different text versions in relation to these stages, something which also entails, at many seats of learning, different submissions to your supervisor or seminar leader:

- Prototype 1:1 *Problematization* (also called problem formulation or *problem statement*), in which you make a brief background description and formulate the problem which the thesis will deal with.
- Prototype 1:2 *Research proposal* (often called *thesis proposal*, TP), in which you develop the problematization after having read up on existing research and in which you describe how you had envisaged arranging your study.

We have chosen to include two prototypes in Phase 1, but the number can be adapted to the rules applicable at your department.

Note that the prototypes you are producing (1:1–1:2) constitute the foundation for the sections to be included in your finished product. Often, your department will have a ready-made *template* regarding how the finished paper must look. Later on in the book we present two examples of such templates. We recommend that, right at the start of the Formulate Phase, you get acquainted with these and embark on a discussion with your supervisor about the type of template that will match your research. In line with our suggestion that you should preserve the uncertainty for as long as possible, it is not necessary to choose a template from day one of

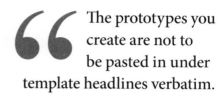

The prototypes you create are not to be pasted in under template headlines verbatim.

the project. The prototypes you create are not to be pasted in under template headlines verbatim. They must be rewritten in order to match the whole. However, you should be aware that the academic paper, in its fundamental features, has a standardized format (even if this format can vary slightly between departments and subject specializations).

Another concrete tip is to be careful about how you *name* the various prototypes so that you can easily find them and differentiate between them. This is particularly important when you are writing with someone else. The best thing is to name the files with the date and then the name of the document, followed by the initials of the one most recently owning and changing it (e.g. "140115 Problematization PB", "140117 Problematization AH"). Then you will easily be able to find the latest version of the respective prototype. Ensuring that all files are constantly available to all authors is self-evident, something which can suitably be done by means of everyone always putting the latest version in a shared folder that everyone can get to. Perhaps your supervisor should also have access to this folder.

Prototype 1:1 Problematization (not just problem-solving!)

The aim of a degree project, as already mentioned, is to formulate a non-trivial and researchable problem. Thus, the first prototype of the 4-phase model consists of a so-called *problematization*.

The problematization (also called problem formulation or *problem statement*) is a brief text of 1–3 pages in which you formulate, from a complicated and woolly set of problems, *a researchable and non-trivial problem*.

Traditionally, engineers and students of engineering are good

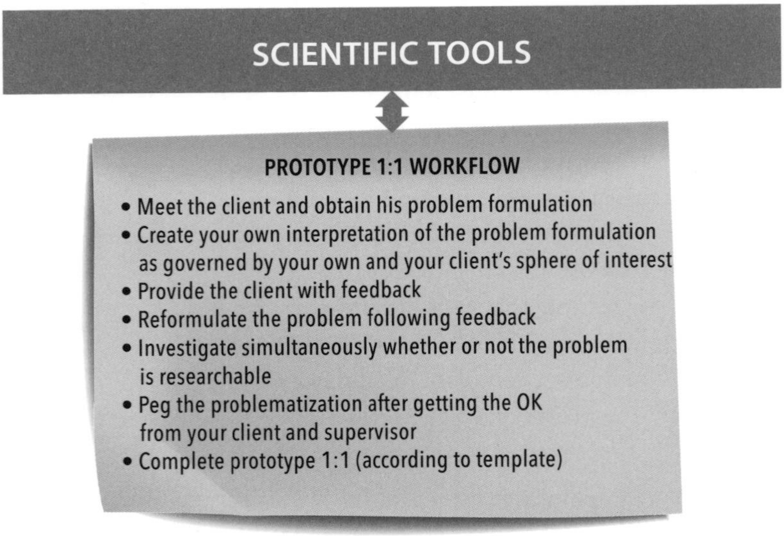

SCIENTIFIC TOOLS

PROTOTYPE 1:1 WORKFLOW
- Meet the client and obtain his problem formulation
- Create your own interpretation of the problem formulation as governed by your own and your client's sphere of interest
- Provide the client with feedback
- Reformulate the problem following feedback
- Investigate simultaneously whether or not the problem is researchable
- Peg the problematization after getting the OK from your client and supervisor
- Complete prototype 1:1 (according to template)

FIGURE 3 Prototype 1:1 Problematization, with proposed workflow. The figure shows that, while work is going on, you need to be aware of all the different scientific tools in order to be able to write Prototype 1:1.

problem-solvers. Technical problems are often dealt with using mathematical models and by designing different forms of optimization models. This ability is, of course, very valuable. However, problem-solving is not enough when it comes to writing a degree project. Neither is a good ability for problem-solving enough when it comes to leading complex industrial change projects, for R&D-management, production, supply chain, logistics, market strategies, organizational change, industrial dynamics, product development, or anything else that a degree project can be about – or that you may end up working with as a graduate engineer. In leading functions in industry and business – as well as when you are writing a degree project concerning these fields – it is a matter of being able to formulate the problem from a complicated context in which many different factors play a part. You have to take into account both quantitative and qualitative factors. Rarely are there any exactly correct or incorrect answers. The industrial context is characterized by complexity and ambiguity. This you will notice when you come into contact with the company or organization, external to the university, that has commissioned you to write your degree project. For these reasons, the main focus is on the first prototype of problematization.

> **To avoid linguistic confusion, we would like to clarify the terminology used in this book:**
>
> 1. We use the term *problematization* when speaking about the increasingly specified and researchable problems that you are working up. The problematization is sometimes called "problem formulation" or "problem statement" and is **not** to be formulated like a question.
> 2. On the basis of the problematization, you will formulate a clear *purpose* of your investigation.
> 3. On the basis of your purpose, you formulate one or more *research questions* which will help you achieve your purpose. These must be precisely formulated but still relatively general, addressing your empirical material as a whole. It is usually said that the research questions operationalize the purpose of a thesis.
> 4. On the basis of the research questions, you gradually draw up *detailed* questions that you use to scrutinize the specific details in your empirical material (e.g. interview-, or survey questions). These questions should not be stated in the introduction to your thesis.

The finished Prototype 1:1 has a simple arrangement (see illustration) whereby you initially account for the background to the problem that the paper will be dealing with (and which is expressed in the problematization). Following that, you clarify which phenomenon your study will be about and what the preliminary purpose of the thesis is. You also formulate a preliminary research question. You should reconcile the problematization with the company/organization you are collaborating with if you have an assignment, and with your supervisor (see Figure 3).

> **Template for Prototype 1:1 Problematization**
>
> - Background (needed in order to understand the client's problem and your own problematization)
> - Problematization (clarifying which phenomenon your study will deal with, which aspect of this phenomenon you intend to study, and why a study like this will be of interest)
> - Preliminary purpose
> - Preliminary research question

WORKING DOCUMENTS

While working on prototyping during Phase 1, you would be wise to gather your impressions and the material you are working in four separate working documents/folders:

- working documents on *scientific legitimacy* (research and writing)
- working documents on *literature* (theory and previous research)
- working documents on *method* (gathering and analysis)
- working documents on *empirics* (source evaluation and references).

In these working documents, you gather reflections, notes, and excerpts concerning each respective field – all in order to develop your insights into your work bit by bit. The four documents can be thoroughly prepared, to varying degrees, depending on how the thesis process looks at your university and on which kinds of texts you are being expected to deliver during the process. The material in the respective document can be summarized, presented, and discussed at seminars or in smaller peer review groups, or used as supportive documentation for meetings with your supervisor. Another area of use for the working documents is as so-called *portfolio examinations*, i.e. when the texts are submitted and form part of the examination, which happens at certain universities.

In the majority of cases, however, the working documents are your own material, which you will continue to build upon bit by bit. They can form the basis for your texts and they work like gathering points for all the material that you collect during the course of your work.

Contents of the four working documents and examples of question formulations to work with

Working documents on scientific legitimacy (research and writing)
- Discuss your research and thesis using concepts that are relevant to you. Position your own contribution in relation to other research in your field.
- Discuss your work in relation to the degree project's learning goals. Use the learning goals as headings and go through how well you meet each individual learning goal.
- Discuss your own writing – address, recipient adaptation, scientific legitimacy.

Working documents on literature (theory and previous research)
- Survey and document the theories, theoretical concepts and findings from previous research that exist in your field and which lie close to your purpose and question formulations. The theory you use can also originate from another field, in which case you will need to especially carefully justify and explain your choice of theory.
- Account for which theory and research fields and which researchers you relate to, continue building on, or challenge, and account for which findings or insights you feel have especially benefitted you.

Working documents on method (gathering and analysis)
- Survey and document the existing scientific methods in your field which match your source material, your purpose, and your questions formulations. Discuss the reliability and validity of these methods.
- Account for which methods and concepts you relate to, continue building on, or challenge, and account for which method focus has especially benefitted you.

Working documents on empirics (source evaluation and references)
- Account for your primary and secondary sources, access, representativeness, and reliability and/or validity by means of a source-critical analysis.

SCIENTIFIC TOOLS
PURPOSE AND QUESTION FORMULATION

It will not be enough for you to choose a subject for your thesis as one subject is almost always too big. One and the same subject can give rise to lots of different investigations, with different purposes.

When finding your purpose, it may be wise to begin with a specific problem that a company or organization has. If you have a client for your degree project, you will be able, within and from the assignment you have been given, to formulate a problem. If you have no client, you can try to imagine whom you would like to have as one and then you can try, by imagining being in his shoes, to find the problem. The next step is to think about what part of the problem can be *generalized* or transferred to other companies, organizations, and contexts.

> Of which general phenomenon is this problem a case?

The key question when finding the purpose, on the basis of an assignment, is thus: Of which general phenomenon is this problem a case?

Often, there are several conceivable answers to the question; whichever answer you choose, i.e. whichever phenomenon you choose to study using the case that the assignment deals with, it can change over time, as you read up on what the research conducted in the field shows, find out more about the case, and so on.

The purpose, besides showing you what is to be done, also shows you what is *not* to be done. The purpose thus *demarcates* the problem you are to investigate. The purpose is also closely linked to the *research design* you choose, i.e. the strategy you choose in order to study what you are interested in knowing more about.

Note that the purpose is not to be expressed as a question. The purpose is a concise formulation of what you are to investigate. It is appropriate to use powerful verbs like *investigate* and *analyse* when formulating the purpose (e.g. "the purpose is to investigate ..."). Do not use woolly expressions like *look at, check up, highlight,* or formulations like "the purpose is to answer the question ...".

A scientific work can have four different types of purpose:

- An *exploratory* purpose entails exploring, using the study you are conducting, something that is unexplored or something that has not previously been scientifically studied to any great degree. Or you are seeking to identify and discover important dimensions of a problem which has hitherto remained unknown or which has not yet been described in the scientific literature. In some cases – e.g. if you have an exploratory purpose in some work that functions as a pre-study for a larger endeavour – you may want to generate hypotheses for further research. An exploratory purpose is often combined with an inductive approach since you are unable to know for sure exactly what you are searching for. An exploratory purpose is very common in social science research.

- A *descriptive* purpose is appropriate when you want to understand more about a phenomenon which has already been demonstrated in previous research, but where knowledge of that phenomenon is limited. This entails wanting, with the aid of the study, to describe a phenomenon, perhaps because the phenomenon is new and has not been documented before.
- An *explanatory* purpose entails wanting, with the aid of the study, to explain the cause and effect, i.e. causal links, e.g. why something looks like it does or why people act in a certain way.
- A *predictive* purpose entails wanting, with the aid of the study, to try to predict the consequences of something; what the effects will be and in what way.

When it comes to social science research, explanatory and predictive purposes are often difficult to fulfil since research like this most frequently concerns dimensions on various levels that interact in a frequently very complicated way.

It is usually said that *the research questions operationalize the purpose*. This means that the questions are to be formulated in such a way that you, as a researcher, using the answers to these, can achieve the purpose of your investigation (see example in Figure 6). During

Purpose of study	Examples of research questions
Exploratory	What happens to X? What are the underlying themes, patterns, categories behind X? How do people perceive X?
Descriptive	What are the underlying behaviours, events, attitudes, structures, processes in X?
Explanatory	Which events, outlooks, attitudes, policies create X?
Predictive	How does X affect Y? In what way?

FIGURE 4 Different types of purposes, with examples of associated research questions.

> **The research question is the question that you ask of your entire empirical material.**

the course of your investigation, you will certainly discover new things that force you to change, and sometimes completely replace, both the purpose and the question formula. For example, your purpose may be to investigate what kinds of engines will be used in the cars of tomorrow. One of your question formulations will then be about electric engines, but after a while, you realize that your purpose has been specified and that you are actually investigating the prerequisites of the electric engine as a future alternative. In this case, one of your question formulations has been converted into the purpose of the investigation, and new research questions thus become relevant. The ultimate purpose of a scientific investigation is actually made clear at the very end of the research process, when the thesis is about to be finished. Not before then will you know with certainty what purpose you had and which questions you have actually answered.

The research question associated with the purpose is not the same as the detailed questions you pose in the study you are doing. It is more a matter of the questions that you, as a degree project author, ask of your entire empirical material. Expressed differently: your empirical material will assist you in answering your research questions.

This means that your research questions have to be written in such a way that you can answer them using your empirical material. When working *inductively* – which entails focusing on empirics before choosing a theory – it is almost impossible to formulate, right from the start, in Prototype 1:1, the right questions. The usual thing, while work is ongoing, is that you will change the questions, and the purpose, when you notice both what is of interest and what your empirical material actually deals with. This is not unscientific at all. Rather, it is a part of the inductive working method of allowing the empirics to generate theory; in order to do that, you have to be open to what the empirics contain and, on that basis, adjust the purpose and the research questions.

> **Checklist for a good purpose and a good question formulation**
>
> - The purpose has a clear (if implicit) recipient (it is evident who can benefit from the findings and whose perspective you want to construct in the study).
> - The question formulation has not been answered previously but can be studied.
> - Both the purpose and the question formulation are based on already-existing knowledge but can also lead to new questions once the purpose has been achieved and the research question has been answered.
> - The terms used in the purpose and the research question have been chosen with care and are coherent with the problematization that precedes the purpose formulation and the research question in the thesis.

On the basis of the purpose of your work, and the associated research question, you choose your *research design*, i.e. the way in which you gather in material as well as the *data gathering method(s)* you will be using. (Read more about research design on p. 62 ff and about data gathering methods on p. 74 ff.)

APPLYING A SYSTEM PERSPECTIVE

The problematization is to include a *non-trivial problem*. By this, we mean a problem containing several cohesive and mutually-influencing levels of the industrial or organizational operation and where the answer is not, therefore, given in advance. It will not be enough to solve the problem on one level since a change on this level will affect other dimensions of your client's company or organization; you will need to apply a *system perspective*.

This entails you thinking about how the problem you have encountered may be understood on three levels (see Figure 5):

- *The individual and organizational level*, i.e. seen from the perspective of management andemployees. Important areas include management and executive philosophies, organizational culture, organizational identity, and HR issues.

> " It will not be enough to solve the problem on one level; you will need to apply a *system perspective*.

- *The functional level*, i.e. seen from a process and production perspective. Relevant dimensions here include organizational structure, supply chain, and logistics.
- *The industrial level*, i.e. seen from a wider industry perspective. This perspective includes dimensions connected with, for example, industrial development and dynamics, regulations, globalization and national/international finances.

After that, you can suggest a more specific problematization that suits both your fields of interest and those of your client.

The system perspective is also relevant when you are *not* writing your degree project on the basis of an assignment given by a company or organization that is external to the university. Maybe you have come across a subject for your degree project after having come across/stumbled over something interesting or puzzling in the media, or after having attended a certain course or talked to your supervisor. Regardless of how you arrived at the subject of your degree project, you should construct a system perspective. And, as previously mentioned, regardless of whether or not you are writing your degree project on the basis of an assignment given by a company, it will still be appropriate for you to identify a client who would be interested in your investigation. This will help you to envisage this client's situation, problems, and challenges.

FIGURE 5 The system perspective involves the problem that your thesis deals with being seen from an individual and organizational perspective, a functional perspective, and an industrial perspective.

> **An engineering company's thoughts on outsourcing**
>
> An engineering company had observed that several competitors had outsourced their production to China during recent years. It tasked two Master's students with investigating whether it, too, should outsource its production or whether it would be wisest to let production remain at its factory in Sweden.
>
> The Master's students began by recollecting what they had learnt during a course on outsourcing. They then sketched out, on the basis of their existing knowledge, and jointly with their client, which dimensions of the outsourcing-problems the company was facing in accordance with the system perspective:
>
> - Individual and organizational level: How is outsourcing implemented? What HR factors need to be taken into account? What intercultural problems can arise during outsourcing?
> - Functional level: What new logistical challenges would arise during outsourcing? What does it cost to shut down a factory?
> - Industrial level: How do the trends look with regard to the outsourcing of production in the engineering industry as a whole? What is happening to the Chinese currency vis-à-vis the Swedish currency right now?
>
> In consultation with the engineering company, the Master's students decided to limit the problems to the logistics issue (functional level). They wrote their first prototype, problematization, where they briefly described the trend within the industry, that the engineering company was also thinking about outsourcing in order to enhance its profitability, that the logistics were the biggest question mark, and that they thus needed to be investigated.

DIVERGENT AND CONVERGENT THINKING

During the Formulate Phase, when you are about to problematize and create a research proposal (Prototype 1:1–1:2), it will be a good idea to switch between divergent and convergent thinking. We illustrate this using a graphic called the recumbent Christmas tree which contains these two recurring elements (see Figure 4).

You should avoid aiming for a certain solution too early. It is very easy to converge (focus) on the first solution. But in this case, there is a risk of getting bogged down in a certain way of thinking. And, in all probability, your first solution idea will probably not be the most

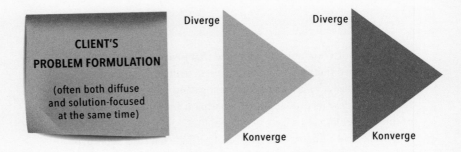

FIGURE 6 Divergent or convergent thinking. The goal is to retain the uncertainty and not lock the research focus too early on by means of switching between diverge and converge, as the case may be.
Source: Adapted from Mats Engwall (ed.) (2004), *Produktutveckling bortom kunskapens gränser: Mot en osäkerhetens grammatik*. Lund: Studentlitteratur.

optimum one. At regular intervals, you should take a step backwards in your thoughts and diverge (broaden) the problematization, purpose, research question etc. in order to thus call into question the direction your investigation has taken.

At the beginning of the research process, it is especially important to switch between divergent and convergent thinking. Eventually, you will have to converge in a certain direction, of course. But do not do this too early on.

Diverging – broadening the problematization

A student was commissioned by a major company that delivers its services via the Internet. The company's problem description looked like this:

> A large part (>80 %) of the incoming cases dealt with by customer services concern recurrent issues, e.g. that people cannot log in. Users should, in many cases, be able to rectify the problem themselves, but today, they contact customer services who must manually help each individual. Our goal is to reduce costs by automating as much of this as possible.

> Immediately, the student started searching through the research literature concerning the problem of automating and routinizing customer services, as well as similar activities used by Internet-based service companies. Very soon, he got stuck. The reason for this was that he got bogged down in the problem formulation that the company had provided because he immediately focused on the kind of solution that they had indicated.
>
> The advice given by the student's supervisor was as follows (and we agree):
>
>> Take a step back. Try to think about what the company's problem actually is. Do not focus too quickly on a solution as this will run the risk of being suboptimal, and you will not be giving the company any more than what it had suspected already. Do not go off, so to speak, on a tangent. Instead, broaden the set of problems. A more general problematization would instead be the company wanting a more cost-effective level of customer satisfaction. In this problematization, the automation of customer services is only one potential solution proposal. Additionally, this will prompt you to search for previous research, theories, and models in the wider field of customer satisfaction. In this way, you will satisfy both the company, by giving it more than what it had suspected already, and your supervisor by means of a clear and scientific grounding of your problematization.

INDEPENDENCE AND MANAGING (DUAL) CLIENTS

It is not unusual for students to write their degree projects on the basis of an assignment they have been given by a company or some other organization external to the university. This company/organization may want assistance in investigating something, producing a model for something, or making some form of recommendation, based on a scientific study. As a degree project author, you are expected not only to carry out the assignment, but also to author a scientific paper in conjunction with this which will become the icing on the cake that is your academic education as a BSc or as an MSc in engineering, or a Bachelor's or

> *The company has the function of assisting you in finding an external problem – the university has the function of helping you develop necessary knowledge and proficiency.*

Master's degree. Often, you will have one supervisor at the company/organization and one at the university.

This means that *degree project* is used as the designation of both the process/product you are making for, and delivering to, the company/organization and of the process/product you are making for, and delivering to, the university. Many students feel that the two clients give them mixed signals regarding what they want. While the company/organization is interested in useful results, and maybe also recommendations on how to act in a certain issue, the university supervisor emphasizes scientific legitimacy and an academic approach.

We are of the opinion, as already mentioned, that these two clients actually do not have such disparate interests as one might imagine. The company/organization and the university are both interested in the assignment being conducted with as high a level of quality as possible.

On the other hand, these two different clients have different functions during your degree project process. While the company or the organization has the function of assisting you in finding an external problem, i.e. a problem which is of interest to someone outside the university, the university's task is to help you to develop the knowledge and proficiency regarding scientific method and thesis-writing that you need in order to carry out the assignment with as high a level of quality as possible. In that, we find everything that this book raises: the significance of continuing to build on previous research into the problem, purpose, and research question, consciously being able to choose the right method for your research design, data gathering, and analysis so that the problem becomes researchable – in brief, providing you with insights into all the dimensions of scientific legitimacy.

> A degree project can thus be *client-driven*, *research-driven,* or self-driven, i.e. driven by your own curiosity.

This book and the work model presented are adapted to a situation where you are going to identify the problem and the question formulations on the basis of the external client's often diffuse problematization. If you do not have an external client – which not all degree project authors have – we thus recommend that you make up a fictional, but authentic, one. Formulating a problem for a client will give your thesis a clear purpose and a focus which will help you when you are writing since you will be able to envisage your recipient's frame of reference and thus know at which level to pitch your language and explanations.

> **Examples of instructions concerning confidentiality issues linked to degree projects**
>
> 1. The company can refuse you access to certain sources or the use of certain facts during your investigation.
> 2. Everything that you actually investigate must be discussable at seminars held within the course.
> 3. Everything you investigate must be reportable and discussable in your report (with the possibility of anonymity).
> 4. Not all details included in your investigation have to be in the finished report. This can, for instance, apply to numbers in a table or to other sensitive data. This balance is something you will have to arrive at in consultation with the company and your supervisor.
> 5. A so-called executive summary is a summary of what is written in the report. This may not include any other information than what is in the report.

SOURCE: INSTRUCTIONS USED ON THE INDUSTRIAL MANAGEMENT MASTER'S PROGRAMME, INDEK, ROYAL INSTITUTE OF TECHNOLOGY, WRITTEN BY LARS UPPVALL AND PÄR BLOMKVIST.

A degree project can thus be *client-driven*, *research-driven*, or *self-driven*, i.e. driven by your own curiosity. All three entail similar problematization work. In all cases, you will have to work just as hard in order to convert the diffuse problem into a researchable one.

One major difference, however, is that it is often easier to relate *impartially* to what you are studying in a research-driven or self-driven degree project vis-à-vis a client-driven project since you do not have any special bonds with the empirical material you are gathering.

In the client-driven degree project process, it is also more difficult to be impartial as you will develop a close relationship with the client and, for natural reasons, you will be interested in receiving good references from him – maybe even a job once you have graduated. This is something that you, as a degree project author, have to be aware of and aspire towards counteracting – the quality of your finished work is dependent upon your remaining impartial while it is on-going. (Please note, however, that impartiality and objectivity are not the same thing.)

You have to be able to write about everything that can be investigated. However, the company can refuse you access to some parts of the organization, or prohibit you from using some information. It is very important for you, as a student, to be aware that academic reports are in the public domain. In consultation with your supervisor and the company, you will have to find out if the restrictions are so strong and the confidentiality so high that it will be impossible to write a degree project. In this case, you will have to decline the assignment, no matter how exciting it may seem.

TOOLS FOR THOUGHT – VARIOUS KNOWLEDGE INTERESTS

A tool for thought, used to understand both your client's and your own way of thinking, involves considering which kind of knowledge interest you have – that is, why you are interested in studying what you want to study. Many engineers, in addition to engineering science in general, are driven by a *technical knowledge interest*, which is not connected with technology in any literal sense. Wanting to understand how a certain management model works in practice in order to be able to improve it is a typical technical knowledge interest. Wanting to understand how managers' behaviour makes it easier/more difficult for women to advance within an organization can be one example of a *historic-hermeneutic knowledge interest*; however, if you additionally hold the opinion that it is wrong for women, in practice, not to have the same career opportunities as men, accompanied by a will to change this, then there will be an *emancipatory knowledge interest*.

> **Habermas' three knowledge interests**
>
> - *the technical knowledge interest*, where the driving force behind the knowledge development lies in explaining how something works, learning more about something in order to be able to use the knowledge to improve, for instance, a process or product
> - *the historic-hermeneutic knowledge interest*, where the driving force behind the knowledge development lies in developing knowledge in order to understand something; making something complicated comprehensible
> - *the emancipatory knowledge interest*, where the driving force for developing knowledge is wanting to understand something for the purpose of changing something deemed (morally or ethically) wrong.

SOURCE: JÜRGEN HABERMAS (1987), *KNOWLEDGE AND HUMAN INTERESTS*. CAMBRIDGE: POLITY.

Thinking about knowledge interests can thus function as a tool for thought and help you to bring order to your purpose. It can also help you to understand the literature you are reading since different theory formations have been created as a result of different researchers' knowledge interests.

ETHICS

Closely linked with the issue of managing (dual) clients and your independence as a researcher are issues regarding research ethics. When you are working with the phases of the degree project process, you are expected to behave in an ethically correct manner. You are also expected, in many cases, to reflect upon how you have done this in a particular section of your paper, often in the method section.

In order to guide both yourself and other researchers towards ethical actions during the course of the research process, there are different codes of behaviour. You should investigate whether or not there are any ethical codes in place at your university; if so, become acquainted with them and think about the ways in which they affect your work.

On the national level, the most common ethical codes within the social sciences in Sweden are the *Swedish Research Council's principles of ethical research for the humanities and social science*. These codes are something that all researchers wanting to study a social science phenomenon must comply with. The Swedish Research Council's paper describes four different principal requirements which scientific work in the social sciences – including degree projects – has to meet:

- *the information requirement*, which entails that the people you are studying (e.g. by interviewing them or asking them to answer a questionnaire) have to be informed about the purpose of the study
- *the consent requirement*, which entails that those you are studying have to agree to be studied
- *the confidentiality requirement*, which entails that the material you collect/create when conducting your investigation has to be treated confidentially (thus, it must not be shared freely and you may need to make companies/organizations anonymous as well as refrain from writing about your informants in a way that makes them identifiable)

- *the good use requirement*, which entails that the material you collect may only be used for the purpose that you have stated when collecting the material.

Ethics are thus about complying with good praxis regarding scientific work so that nobody will be harmed by your work. However, ethics are also about adhering to the norms of scientific work when you are presenting your work in spoken and written form so that others can see that you have remained impartial during the process, and so that others can evaluate your findings. It is thus of major importance that you make correct references to the different kinds of *sources* you use in your work. Your readers (or listeners) must themselves be able to determine how and to what extent you are continuing to build on existing knowledge and what originates from you. In order to be able to do that, it is very important that you specify when you are quoting or referring. *Quotes* are to be reproduced verbatim and demarcated using inverted commas, while *summaries* are to be your own paraphrasing (i.e. using your own words) of something you have read. In both cases, you have to specify the source (in accordance with the reference management system you are complying with). When you quote, you must also specify the page on which you found the quote in the source you are quoting. (Read more about sources and reference management on p. 122 ff.)

> Ethics are about complying with good praxis regarding scientific work.

If you break the rules regarding quotes and summaries in a way that makes it seem that you yourself have discovered something that you actually read somewhere else, you may be accused of *plagiarism*, which is a very serious offence against scientific ethics and something which, at many universities, can lead to your being sent down. Note that, in this book, and a tad contradictorily, we do not use a reference regime that is equally as strict as the one we advocate for the academic paper. The reason is that we now comply with the praxis applicable to the genre of textbooks.

> **Plagiarism**
>
> A person doing a degree project was commissioned by a major American carmaker to investigate the Chinese market. It was a rush job. The company wanted quick results and the person was promised a job immediately after graduating. At the final seminar, the examiner discovered that some parts of the paper had been written in perfect academic English. Other parts had not been badly written but did not at all achieve the same level. A few simple searches revealed that large chunks of the theory section had been pasted into the paper from other texts. The supervisor from the carmaker did not think this was strange at all and pushed for a quick pass. The student confessed but did not feel that this was of any major significance. The examiner was forced to fail the project and the student's degree was delayed by several months.

LEARNING GOALS AS A CHECKLIST

Our experience is that students rarely have any major interest in syllabuses and learning goals. We recommend, however, that you become acquainted with the learning goals as they are set forth in the syllabus your thesis course so that you know what you are expected to know when you have finished your paper. It is, of course, these learning goals that the examiner will assess, once the thesis is finished, in terms of whether or not you have achieved them. The learning goals are thus connected intimately with your grade. You can also make use of the learning goals (as well as any specifications of what the degree project has to show in order to attain a certain grade) as a checklist during the course of your work to ensure that what you are doing will lead to the achievement of your goals.

Prototype 1:2 Research proposal – explicitly defining and making your problematization scientific

The initial problematization now has to be explicitly defined and formulated as a scientific problem. We call this Prototype 1:2 Research proposal. At many universities, this is also called a *thesis proposal* (TP), and it is common to hold a seminar at which the research is presented to supervisors, seminar leaders, and fellow students.

The problem you formulated in Prototype 1:1 was, in all probability, dependent upon the specific situation your client is in, with the solution to this problem perhaps being of greatest interest to the client rather than to a broader circle of stakeholders. The same will probably apply even if your problematization originates from your own or from you supervisor's ideas. If this is the case, your investigation risks becoming trivial from a scientific perspective. In order to reformulate the problem into a non-trivial one, which is of broader scientific interest, it will have to be reformulated so that it becomes "new" in some way in relation to existing research in the field. This means that you will have to begin reading up on previous research and, while you are reading, you will have to think about how the problem you formulated before, as well as a study of this problem, could contribute towards existing research into the field. It is important to note that existing knowledge of the field means the knowledge gathered in the scientific literature, and *not* the knowledge that you or your client possess, or the knowledge of an individual expert who has not written about his/her knowledge in a scientific way.

> You are expected to contribute something more and something new, compared with what exists in the previously assembled research.

You are expected to contribute something more and something new, compared with what exists in previously assembled research.

Scientific knowledge can entail that you make a theoretical, methodological, analytical, and/or empirical contribution.

It is not particularly common for degree projects to make any major *theoretical contributions* – there is simply not enough time to do a project of the extent and depth that this will most frequently require. Above all, there is seldom time for reflection in the manner most frequently required by the development of a theoretical contribution. However, there is nothing to say that you cannot aspire towards developing an existing theory on the basis of your study.

Some degree projects make *methodological contributions* by showing new ways of solving an organization's practical problems. A common way of doing this is by gathering *best practices*, i.e. examples of good ways of doing things, and to use them in order to further develop or refine the existing literature.

Other degree projects make *analytical contributions* whereby students, by analysing a problem from a new angle – perhaps by combining theories from two different but compatible theory fields – show new ways of understanding empirical data.

In our experience, it is very common for degree projects to primarily make an *empirical contribution*. Here, the degree project quite simply sheds light on some new, i.e. in existing research a hitherto not particularly well-documented type of, empirics. This can, for instance, occur by means of a description of a new type of empirical question, another type of company/organization than is usually studied in the special field, or an entirely new empirical phenomenon, e.g. a new process or a new concept.

> " You will in all certainty, during your journey, change what you have written in Prototype 1:2 several times.

The most important question you have to ask yourself when working with Prototype 1:2 is whether or not what you are presenting is researchable and non-trivial, and whether or not all parts of the prototype are logically coherent (see Figure 7).

However, and this is important, you will in all certainty, during your journey, change what you have written in Prototype 1:2 several times. Despite this, it is important to determine, early on, the focus of your degree project. Then you will know why you are changing your mind and be able to justify your choices and demarcations. The pains you have gone to in order to prototype a research proposal will be

SCIENTIFIC TOOLS

PROTOTYPE 1:2 WORKFLOW

- Start out from Prototype 1:1
- Read up on the scientific tools (e.g. theory, previous research, method) in order to obtain a clearer picture of your research field, your purpose, and your question formulations
- Reformulate, if necessary, the problematization after reading up
- Survey your empirics and which gathering methods are suitable; if necessary, do a "pilot study"
- Complete Prototype 1:2 (according to template)

FIGURE 7 Prototype 1:2. Research proposal, with suggested workflow. This figure, too, is aimed at illustrating the iterations between the scientific tools in the various elements needing to be worked through in order to formulate the prototype.

far from in vain, even though your thesis, over time, may take off in another direction.

The finished Prototype 1:2 is more extensive than the previous prototype and continues building on it by means of a revised background description, problematization, purpose, and question formulation (on the basis of what you learnt when reading up on the field). In this prototype, you should additionally be able to clarify what your study will be able to contribute in relation to existing research in the field (positioning); you should be able to describe the literature and theory you believe to be suitable for the study (literature/theory), and maybe even to conduct a simple line of reasoning as regards how you envisage conducting the empirical study (method).

This prototype is longer than the previous one. Now (as in later prototypes), it will be important to comply with the requirements governing academic work with regard to formalities. Do not forget to number the pages of your document, and to make sure that you quote/summarize in a correct manner. You should also include a reference list.

Template for Prototype 1:2 Research proposal

- Background
- Problematization
- Purpose
- Research question(s)
- Your study's expected contribution – the positioning of your investigation in relation to previous research
- Literature and theory – the literature you believe may be suitable, the (theoretical) concepts you believe you may need in order to understand the material you are gathering in for your study
- Method – a description of how you can acquire empirical material for the investigation
- Reference list

SCIENTIFIC TOOLS

SEARCHING AND RELATING TO LITERATURE

Authoring a degree project entails conducting scientific work. Since science entails searching for fresh knowledge, it is natural to start by finding out what has already been done as regards the phenomenon you are interested in. What knowledge exists already?

The answer to that question will be obtained when searching the literature within the field. By literature, we mean all forms of published material. This can include books, journals (both paper and digital), student projects, conference papers, and so on.

Finding literature on what you want to study can sometimes seem difficult. You either find too little or too much. If you are writing a thesis based on an assignment given by a company or another external party, the phenomenon you are interested in will often be so complex that it points in many different directions, providing you with many different fields to read up on. A certain amount of *über reading up* will often be necessary.

By über reading up, we mean that you, as the author and researcher, will have to read up broadly in order to understand what you are going to do yourself. Maybe you will have to read or orientate yourself on fifty scientific articles even though you will only directly benefit from ten. The problem areas you are facing are often so diffuse and complex; additionally, the subject and discipline boundaries in our scientific field are not at all as clear as those in, for example, natural science. You can find literature relevant to your project in such disparate fields as organization theory, psychology, the history of economics, and management theory. We are working with a kind of interdisciplinary science and thus we have to overview a large area.

One way of starting to search for literature is to look at textbooks which seem relevant to what you are interested in. This is a good start as textbooks often contain a basic overview of the field that the book deals with. Additionally, textbooks often contain reference lists which you can use to progress in your literature search. Once you have found new literature, you can use the reference lists in new texts, which will thus help you along. After a while, you might be able to identify specific scientific journals, authors, and researcher groups that you will then be able to search on in order to find further material.

Often, the librarians at the university library that you have access to will be able to help you get started with your literature search. The

> **IT in public administration or expectation management during processes of change?**
>
> A student was in contact with a government agency that was about to develop and introduce a new IT system. As she was not sure what her thesis was going to be about, she started reading up broadly: she read about eGovernment, knowledge-intensive organizations, change management, and business development. Her first problematization (Prototype 1:1) dealt with the introduction of IT into a knowledge-intensive organization. However, in step with her increased reading, and conducting talks with her points of contact at the agency, she realized that what was of most interest to her, both in relation to what had previously been written and what the agency was interested in knowing, was expectation management during the process of change. What she had read about eGovernment and IT development/implementation then became comparatively irrelevant in relation to her new problematization; however, at the same time, reading was necessary in order to realize that this was what she would *not* be writing about.

university libraries have access to both their own and other databases where different forms of literature may be found. To navigate right, you will need to define search terms for your work. You may need to change these once you have got started, as you read up and define your research questions.

Reading up on a field and doing a literature study thus entails identifying what has been published in the field, reading up on that, and then summarizing what you have read in an argumentative text.

The literature study works as a way of generating ideas regarding what you are going to write about; in addition, it is a way of refining the ideas that you already have since the literature review shows what knowledge (generally) exists in the field. By reading up on what has already been done, and by showing this in your paper, you will also be showing how you are continuing to build on existing knowledge within the field, which is an important part of doing science. This is called *positioning* your own study in relation to previous research in the field. Questions you can ask yourself while positioning include:

> " **You will have to read broadly in order to understand what you are going to do yourself.**

- What will the contribution made by my study be?
- What knowledge gap(s) will my study fill?
- In what way does my study involve another perspective vis-à-vis existing studies?
- Why is my study needed?

> Reading up on a field entails identifying what has been published in the field and then summarizing it

When reading, you have to have a *critical mindset*. Being critical in this context entails having a questioning attitude towards what you read. (Read more on critical thinking on p. 54 ff.) In concrete terms, this means continuously scrutinizing the methods used to deliver previous knowledge, the findings presented, and the conclusions reached. Additionally, you should also be attentive to which assumptions underpin the literature you are reading and you should think about whether or not these are reasonable, or if you can see things in another way. We will return to this in the section on source evaluation and criticism on p. 124 ff.

Contents of the literature section

- a description of the problems that previous research has tried to explain or understand explanations of the theories, concepts, and models that have been used
- critical argumentation regarding why existing literature is not sufficient, or what it has not yet succeeded in explaining.

In the literature section, you show that you know what previous research has said about the phenomenon you are studying and how your own work supplements and/or criticizes the existing literature.

It is important to remember that the literature study you are writing is to be an adapted description of existing knowledge in the field, *not* a point-by-point summary of each article and book you have read. The literature section is to be written in essay form and have a structure that matches your own work. Content-wise, you have to demonstrate your critical attitude, not just by neutrally accounting for what others have written about the phenomenon you wish to study, but also by illuminating potential weaknesses (to the extent that these are relevant to your work), potential limitations, and so on. Needless to say, in the literature study, too, you will have to comply with the existing rules regarding quotations, summaries, and reference management. If you do not do this, you may be accused of plagiarism (see the section on ethics on p. 37 ff.).

WHAT IS THEORY?

The literature you read can vary in terms of being theoretical. Daily newspapers and business journals are examples of sources that are frequently more phenomenon-describing than theoretical. Scientific articles that have been subjected to the *peer review* process (being reviewed by other researchers, leading to a more or less extensive reworking of the text prior to publishing), are often more theoretical than phenomenon-describing.

Theoretical means that the text attempts to summarize the knowledge of a certain phenomenon in a general way in a proposition that is called theory. All research ultimately aims to develop theory about the phenomenon under study, even though intensive discussions – primarily over the last half-century – have taken place in the scientific community regarding whether or not this is possible. It has, for instance, been asserted that it is impossible to posit a general proposition in the form of a theory on reality as it is considerably more complex than that.

Most social scientists, however, agree that social science theories do not tell the truth about the world; instead, they are models, or thought tools, that we can use to understand different phenomena. "All models are wrong, but some can be used", is something the British statistician George Box is supposed to have said. Social science theories can thus neither be confirmed (verified), in the strict sense, nor toppled (falsified) by means of your discovering an exception through your empirical study. They can, on the other hand, be developed or challenged through the development of another theory, or supplemented through the development of a theory that views the phenomenon from another angle.

In the scientific work that constitutes a degree project, one or more theories can either create understanding of, and/or explain, the phenomenon that has been studied and described. This type of work, *phenomenon-driven work*, entails starting out from a knowledge gap that can be found in the existing literature regarding an industry, a new type of working method, an organizational form, and so on. The theory can also form the starting point for problematizing the work, when a knowledge gap has been identified between two theories, or a lack of understanding of a certain perspective in a theory. In this case, the work is *theory-driven* and aimed primarily at developing theory rather than explaining a certain phenomenon.

Regardless of whether the work is phenomenon-driven or theory-driven, the theory and *the theoretical concepts* will help you develop an understanding of what you are studying. You use the concepts when describing and explaining what you have investigated in your study. It is these theoretical concepts that constitute your theoretical framework. In the theory section of your degree project, you will present the theories you are using to fulfil your purpose and answer your questions. You will not be accounting for all imaginable theories that can exist in your scientific field and neither will you be writing all you know about the theories you are using.

> Theoretical means that the text attempts to summarize the knowledge of a certain phenomenon in a general way

Theories in social science research

- are often adapted to understanding/explaining empirical phenomena of limited extent (e.g. how organizations function, how an industry changes, how a company's value chain works, how learning comes about in organizations, and how stakeholders are involved in an organizational process)
- are most often not used to discover general natural laws or strict conformity to laws, which also means that they are rarely used to predict events
- are used to generalize and summarize knowledge of a certain phenomenon in a general way
- are used when comparing and discussing findings within a scientific field in relation to other findings within the same field
- provide perspective and thus work as lenses, thought models, or thought tools, which can be used to understand different phenomena
- clarify the difference between connections and cause and effect; finding a connection between two phenomena or events is not the same as having established causality, i.e. a cause–effect relationship (for example, there is a clear connection, albeit not a causal one, between the mean temperature of our asphalted roads and the number of drowning accidents).

RELATIONSHIP BETWEEN THEORY AND EMPIRICS: INDUCTION, DEDUCTION, ABDUCTION

You can make use of theory in chiefly two different ways when carrying out scientific work. Literature either works as a means of identifying theories and ideas which are then put to the test by means of conducting an empirical study. Here, you formulate *hypotheses* on the basis of theories and then design a study to see whether these hypotheses can either be verified or falsified. If the theory fulfils the function of this work, you will have a *deductive* approach. When working deductively, a lot of time, when you start working on the thesis, goes into reading up on and creating reasonable hypotheses in relation to existing theory.

If, on the other hand, you are conducting an empirical study on the basis of the problem you have identified, subsequently making use of the theory in order to develop a better understanding of the findings, you will be using an *inductive* approach. The inductive approach thus entails being prepared for the fact that the empirical findings might possibly lead to another theoretical framework than the one you started with; it is the empirical material that shows which theory is of interest.

Sometimes, there is also talk of *abduction* when describing how to switch between theories and ideas in the literature and empirical material being studied, and how the way in which we read the literature is influenced by our understanding of the empirical material and vice versa. The strength of this spiral movement lies in its involving a great level of sensitivity to the empirical material. When working inductively and abductively, however, analysis takes a long time because we often turn to new literature and read up on new theories in order to understand the findings from a study.

SCIENTIFIC WRITING

Scientific texts constitute a *genre* of their own. This means that scientific texts differ from other types of texts, e.g. newspapers or fiction, with regard to form, style, and language.

Even though the *form* of various scientific texts can look slightly different due to the kind of scientific text being written, all scientific texts (e.g. papers, dissertations, conference contributions, and scientific articles) comply with approximately the same tripartite arrangement. There is an introduction, a discussion, and a conclusion:

PHASE 1 FORMULATE

Part 1: Describe what you are thinking of doing
Part 2: Do what you said you would do
Part 3: Describe and discuss what you did

A common mistake made by students is to relate, in chronological order, everything they have thought of during the process and exactly how they have arrived at their conclusions. But that does not work well. The reader is interested in your findings and wants to see that they are well substantiated in previous research and in your empirical investigations. A comparison here could be directions that you give to someone who asks. The recipient is not primarily interested in hearing about all the times when you yourself took a wrong turning, or the obstacles you encountered when looking for the best route. The recipient wants an effective description of how best to get from A to B. Thus, writing the academic paper is like rearranging the description of the degree project process in accordance with the rules that govern the genre of scientific texts. You have to ensure that you only include the information the reader needs in order to understand your research focus, and in order to be able to assess the credibility of your findings.

As regards scientific *style,* the social sciences (and humanities) often differ from technology and the natural sciences. While scientific texts in technology and the natural sciences are often characterized by abstraction, an entirely invisible author (the word *I* most frequently being avoided), and passive voice constructions (e.g. "in this thesis, a deductive approach has been used"), scientific texts in the social sciences are often more essayistic, i.e. closer to the pop science style. An essayistic style means that the text has a line of reasoning and that the author conducts a dialogue with the reader. *I* or *we* are used (when these words are needed; e.g. to describe choices made regarding method), and the language (see below) should not be too high-flown. The advantage of the essayistic style is that it is more reader-friendly; we recommend that you write your degree project in accordance with a style like this if your thesis, subject-wise, concerns the field of social science.

However, even though the social science paper is written in an essayistic style, this does not mean that the language should bcome too poetic. Words and terms have to be chosen with care so that they are as precise as possible, which can involve using complicated technical terms as well as everyday words. Clichés (which are common in many of the social-science-oriented fields which you, as a student attending an institute of technology, can encounter when working on

Good scientific style on various levels in scientific texts

	Good scientific style	Common errors
1. Organization as a whole, sections, paragraphs, sentences	Information systematically gathered according to the principle "from the general to the specific". Paragraphs start with a kernel sentence, possibly preceded by a meta text.	Content not systematically presented. Unclear paragraph division. Unclear links between paragraphs. No kernel sentence.
2. Continuity, coherence	Information structure: Sentence structure follows pattern: topic (first) – comment (last). Linking words exist which connect up the sentences: *therefore, additionally, despite, further, on the one hand … on the other, both … and.* Meta text signals the organization of the text or creates a more accessible content: *Similar conclusions can be drawn by …* and explain how different ideas are connected: *In contrast to X, Y has …*	Illogical information structure confuses the reader. Linking words missing (partly). Text is difficult to read, facts stacked up one on top of the other; (partly) lacking linguistic markers regarding use of the information, as well as its aims and relationships.
3. Summary technique	Summaries of sources a) are of relevance to the purpose and the approach to the problem, i.e. propel the subject forwards and support the arguments b) are formulated using your own words c) have summary markers d) are well-integrated into the context.	The summaries a) have an unclear purpose b) are too close to the original source c) lack summary markers. Facts stacked up without clear connections.
4. Concentration	Text formally in written language form with tight language use.	Colloquial (filler, superfluous words).
5. Precision	Word choice well-judged, objective and unambiguous.	Vague expressions, value words, too personal.
6. Grammatical correctness	Text contains complete sentences, correct grammar and punctuation.	Subject/verb lacking, incorrect sentence construction/ referencing and errors of form.

SOURCE: INGA SJÖDIN AND MARGARETA OLOFSSON, SPRÅKVERKSTADEN, KAROLINSKA INSTITUTETS BIBLIOTEK, 2010

your degree project and when talking to people external to the academy) should be avoided, as should colloquial expressions (fillers and superfluous words).

In contrast to, for example, mathematicians, social scientists work with natural language as their tool. We do not have access to a universal and synthetic language which everyone in the scientific community agrees upon and understands. Mathematical formulae are based on such a synthetic use of language in which all terms and concepts are clearly defined. In the social scientists' natural language, on the other hand, almost all definitions are disputed. What do we mean, for instance, by terms like leader, manager, organization, revolution, reform, and politics? It is the researchers themselves who have to define their terms; thus, it is extra important to be systematic, clear, and open.

Note that you do not need to write finished scientific text from your first day of working. At the beginning, and during a large part of the process, it is wise to use *reflective writing* rather than *presentative*

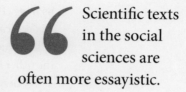

> Scientific texts in the social sciences are often more essayistic.

Tips for reflective writing

- Write ingresses and "postcards", i.e. brief summaries of parts or all of your degree project in which you highlight the most significant aspects. Work with these continuously throughout the entire research process.
- Write/give short talks or presentations in which you tell your audience about different aspects of your degree project. Being forced to express yourself orally to a listener (face-to-face), hones your arguments and reasoning.
- Make storyboards. This is a method used in the theatre and in films to sketch out the most important scenes in order to gain an understanding of the flow and dramaturgy, and to discover gaps of logic in the story.
- Once you have got going, it will be a good thing to make a habit of putting everything you have written or noted down in its working folders on a table or on the floor (a variant of the storyboard technique). Move different parts around physically and think about, or preferably discuss with a colleague, the order in which the different components should be and whether any important constituents are still missing.

writing. The above-mentioned – what characterizes the scientific genre – is typical of presentative writing, to be used in texts which others, e.g. fellow students or supervisors, are going to read. Reflective writing, on the other hand, is what you use to think out loud concerning the process.

When you write reflectively, you should not let the requirements accompanying a certain genre limit you. Instead, you should freely write down your thoughts and reflections, even if this only results in fragments of text. A good way to organise these is to gather them in four working documents in the way described on p. 24.

WHAT IS SCIENCE AND HOW DO WE MEASURE SCIENTIFIC QUALITY?

Your degree project involves conducting scientific work. Science can be defined as a systematic, independent, and critical search for fresh knowledge on the basis of a problematization. To this, we can add the fact that your completed scientific work, the thesis, has to be *logically consistent*. In more detail, these requirements entail:

- making conscious choices regarding issues of method – you should work *systematically*
- remaining neutral as regards what you are studying and critical as regards the literature you are reading and the theories you are using – you are to be *impartial* (see the section on critical thinking below)
- aspiring to understand the deeper significance of what you are studying – you should work *critically* (see the section on critical thinking below and the section "On the analysis of empirics" on p. 114 ff.)
- formulating a problem that leads to your defining a purpose and a question formulation – you must *problematize* (see the section on problematization on p. 21 ff.).

These four criteria regarding scientific legitimacy must accompany you throughout all phases of your degree project – feel free to use them repeatedly as a checklist during the process.

The various criteria regarding what constitutes scientific work are closely linked with the quality of the scientific work. When assessing the quality of scientific work, we speak of *validity* and *reliability*. Put simply, validity entails studying the right thing while reliability entails studying it in the right way. High reliability does not guarantee high validity. But high validity, on the other hand, pre-requires high reliability.

> **Validity and reliability**
>
> *Validity is achieved by means of*
> - the literature review being about the subject field that the problematization, purpose, and question formulations specify
> - the theory being referred to in the theory section, and then used in the analysis, tallying with the purpose and the question formulation
> - the problematization, purpose, and questions tallying with the choice of data gathering method, informants etc.
> - the discussion actually being about what the purpose says it will be about; that the questions are answered.
>
> *Two examples of reliability:*
> - arithmetic reliability: reliability is measured mechanically after agreement is reached between the observers
> - dialogical reliability: there is unambiguity in the interpretation of the empirical material. Demands impartiality and mutual respect on the part of those doing the interpretation.

In all science, scientific work must be open to critical scrutiny by other stakeholders. Fundamental to academia, and a key dimension of science generally, is the *opponent–respondent notion*.

This means that an opponent critically reviews the work in question as it develops and once it is completed. The respondent, i.e. the researcher who has conducted the study and is presenting the findings, is thus given the opportunity to explain, clarify, and be influenced. The same notions form the basis of the *peer review method*. This entails the scientific findings being assessed and evaluated by expert colleagues within the scientific field. As we have already said, we challenge you as an author to put together your own peer review group using your fellow students, and/or some form of reference group or advisory board.

> Science can be defined as a systematic, independent, and critical search for fresh knowledge on the basis of a problematization.

CRITICAL THINKING – THE BASIS OF SCIENCE

The basis of all science is critical thinking. In everyday speech, criticism often entails pointing out errors and shortcomings. In the academic context, however, the word is used in another, more primordial sense. Here, criticism means explaining, interpreting, and aspiring to understand the deeper significance of what is being criticized, or critically reviewed, as it is also termed.

A critical attitude is of key importance to science and is also seen as ethically correct. A critic or opponent must have in-depth knowledge of the subject he/she is criticizing, and the aim must be to make the work in question better. This also means that a critic must not home in on cheap points and obvious mistakes. The critic must highlight more fundamentally unclear points in the argumentation, with the aim of intensifying the analysis.

For you as a degree project author, this critical attitude entails being open as regards where your study is leading you. If you are doing an *inductive study*, it will be self-evident to you that you are continuously reviewing your problematization, your purpose, and your question formulations, not just throughout Phase 1, the formulate phase, but also throughout Phase 2, the design phase, when your work may well make you think again as you will then be seeing the whole soberly. If you are doing a *deductive study*, you will be more locked into the hypotheses you formulated at the start, on the basis of which you carried out your empirical study; however, here too, you will have to adopt a critical approach and be open to reflections surrounding your procedure and the findings from the study. In both cases, it is expected of you that your critical attitude is noticeable in what you write.

> Validity entails studying the right thing while reliability entails studying it in the right way.

In the *literature/theory section*, your critical attitude is noticeable in that you do not just line up one source after another, giving an account of what they have written, but that you arrange them and routinely use a reflective style of writing. It must be clear that you have been reading carefully and that you have been questioning what you have read. In the *method section*, your critical attitude is noticeable in that you justify why you have chosen to arrange the study as you did regarding research design, data gathering method, and analysis method, arguing that your choices are suitable in relation to the purpose of the thesis. Even if you do not expressly account for the methods you have *not* chosen, your critical attitude is also noticeable in that you show awareness of the limitations to the method choices you have made.

In the *findings section,* your critical attitude is noticeable in that you impartially relate to what you have been studying. You avoid value judgements and do not take any particular party's perspective. This also means that you are aware of your use of language, that you avoid clichés and buzzwords (which are very common in companies and organizations), and that you do not take what your informants say as given truths, instead remaining source-critical.

And, throughout the *entire thesis,* your critical attitude is noticeable by means of clear indications regarding what your own comments are and what you have taken from various sources.

> Criticism means explaining, interpreting, and aspiring to understand the deeper significance of what you are criticizing.

Phase 2 Design

It is during Phase 2, *Design*, that you will have to decide in earnest how to design your study and subsequently carry it out. The bulk of your time will be spent on gathering empirical material and compiling it. The objective of work during Phase 2 is to create a report (Prototype 2:1) in which you preliminarily account for the findings from the study you are conducting (see Figure 8). Often, the seminar at which you are expected both to be finished with

> **During this phase, you will need to get involved in the scientific tools.**

and to present this prototype is roughly in the middle of the term, which is why Prototype 2:1 is sometimes called the mid-term report, or *midterm*. In Prototype 2:1, you will be assembling what you have done

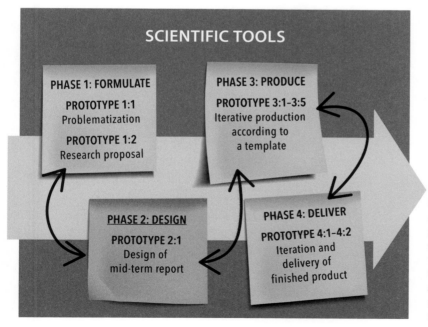

FIGURE 8 **Phase 2: Design**. During this phase, you design a new prototype based on the study you are conducting, the prototypes you have from Phase 1, as well as a more detailed reading up on the scientific tools relevant to your work.

in accordance with the academic paper templates. This means that, during this phase, you will need to get involved, in even more detail, in the scientific tools that are specifically relevant to your work. You need to understand what scientific method entails in order that you may be able to choose the most suitable method regarding research design, data gathering, and analysis in relation to your purpose.

If you are using a deductive approach, you will need to continue reading up on previous research in the field that you wish to study in order to formulate appropriate hypotheses. You will also need to spend a lot of time thinking about how to best test these. Not until then will you be able to conduct the empirical study. If you are using an inductive or abductive approach, you will need to get going with the empirical study and read up on what previous research has shown, in parallel with gathering empirics.

One might easily imagine that the number of deliverables during this phase increases. It may, for instance, be justifiable to complete individual reports on the literature and theory you are thinking of using (a literature study), and on the method choices you are making (a method report). In such cases, Phase 2 would contain several different prototypes/texts. What applies to your university specifically is something that you will naturally need to find out.

FIGURE 9 Prototype 2:1. Mid-term report, with proposed workflow. In order to be able to write this prototype, you will need to implement your empirical investigation. As during previous phases, the production of the prototype is additionally based on your becoming even more deeply engrossed in things that are relevant to your work, and that you feed your newly-gained insights into your work.

SCIENTIFIC TOOLS

PROTOTYPE 2:1 WORKFLOW

- Start out from Prototypes 1:1 and 1:2
- Continue reading up on the scientific tools
- Provide continuous feedback from your reading up and reformulate any problematization, purpose etc.
- Do your empirical investigation and feed back to insights gained from reading up
- Review your plan for empirics gathering
- Complete Prototype 2:1 (according to template)

When working with Prototype 2:1, you may greatly benefit from working actively with *feedback loops*. Feedback loops occur when you relate what you are doing to the reading you are continuing to do on your phenomenon (what you are studying). They also occur when you relate the insights you obtain when reading up on various methodical possibilities, when you obtain the initial impressions of your empirical study, and when you obtain feedback from those around you regarding what you are doing. By writing reflectively in the manner previously described, you will be gathering thoughts and ideas, and thus written material that you can use when sitting down to complete Prototype 2:1.

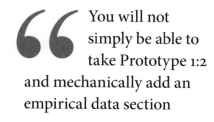

> You will not simply be able to take Prototype 1:2 and mechanically add an empirical data section

Prototype 2:1 Mid-term report

You will need – on the basis of your newly-acquired insights – to re-write the parts included in the introduction to the academic paper (background, problematization, purpose, research question) even if you already have text to start out from in the previous prototype. Thus, you will not simply be able to take Prototype 1:2 and mechanically add an empirics section wherein you account for the findings from your empirical study. The fact is that it is often better to write an entirely new piece of text under these headings in order not to get bogged down in old formulations and thought patterns.

Depending on demands made by your supervisor, Prototype 2:1 might possibly be expected to contain an analysis section, after the empirics account, in which you simply (perhaps in bullet points) account for which method you had envisaged using in order to analyse the empirical material, preliminary thoughts surrounding what the empirical material shows, and how your research questions may possibly be answered.

Prototype 2:1 is expected, just as Prototype 1:2, to follow the standard requirements placed on academic work with regards to matters of form; you should thus have a list of contents, the pages should be numbered, and you should refer correctly. There should also be a reference list at the end.

Template for Prototype 2:1 Mid-term report

Note that, regardless of which academic paper template you choose, Prototype 2:1 must at least have the following content:

- Introduction:
  - Background
  - Problematization
  - Purpose
  - Research question(s)
- Literature and theory
- Method
- Empirics (account of your empirical material)
- (Possible analysis, maybe in bullet points)
- References

SCIENTIFIC TOOLS

ON QUALITATIVE AND QUANTITATIVE METHOD

Method (which derives from the Greek word *methodos*, meaning "pursuit, following after") is the route you take to investigate what you seek to investigate, i.e. the approach you choose when studying the phenomenon you are interested in.

In method literature of different kinds, a division is often made between *qualitative method* and *quantitative method*. As examples of qualitative methods, data gathering methods such as (qualitative) interviews and (participating) observations are usually mentioned, while among the quantitative methods, questionnaire studies, experiments, and statistical methods are usually put forward. The qualitative and quantitative methods are associated with different fields and concepts, as illustrated in Figure 10.

We have chosen, however, not to follow the division into quantitative and qualitative method in this book. One reason for this is that we feel that you, as a degree project author, should not decide too early on whether you will be using qualitative or quantitative method. The field you will be investigating is often difficult to define and delimit at the start of the process. Retain your *inquisitive unclarity* and allow it to also include your choice of method (and naturally also your choice of theories). Additionally, the book's clear process, craftsmanship, and product focus mean that a strict division into two method blocks would work contrary to our purpose. We want you as a degree project author to gradually seek out suitable methods of gathering in and analysing empirical material. The iterative character of prototyping runs the risk, quite simply, of getting lost if you once and for all choose a method during an early phase. Over and above this, one might actually question

Quantitative method	Qualitative method
Numbers	Words
Distance	Proximity
Structured empirical data gathering	Semi-structured empirical data gathering
Deduction	Induction
Generalizations	Contextual understanding
Hard, reliable data	Soft, rich data

FIGURE 10 Common associations with quantitative and qualitative methods.

whether or not it is possible to talk about quantitative and qualitative method; the methods themselves, of course, will hardly be quantitative or qualitative just because the empirical material is. Instead, we have chosen to arrange the rest of the chapter according to the process that you will undergo when working on the design of your degree project. First, you will have to choose a *research design* which is suitable for both your purpose and your question formulation. (Here, however, we are speaking– as you will see – about quantitative studies for the simple reason that the research design we are describing under this heading generates empirical data in the form of numbers.) After that, you choose a *data gathering method*.

Regardless of which method you choose, it is presupposed that you will be dealing with the method issues, when conducting your scientific work, in a *systematic* way. This means that you will consciously choose your research design and data gathering method and that you will be able to justify your choice. It also means that you will consciously choose an analysis method, which you can read about in the section "On the analysis of empirics" on p. 114 ff.

CHOICE OF RESEARCH DESIGN

The research design is a model of how to make the problematization *researchable*. Put simply, the research design entails finding an answer to the question of who or what has to be studied in order to enable you to obtain the supportive data that will solve the problem that the problematization entails.

Thus, choosing a research design concerns thinking about what type of empirical material (*explanans*) will help you understand a certain phenomenon (*explanandum*). Lots of mistakes can be avoided if you differentiate between these terms and return to the question of whether or not you are really studying the right thing, and whether or not you are using the right material to gain answers to your questions. If you are interested in explaining why Lean Management became such a common management model in healthcare (*explanandum*), you will need to gather empirical material that will be useful for finding that out (*explanans*). Maybe you will choose to do a case study of a hospital where Lean Management has been implemented, or you choose to do a survey study in which,

> Choosing a research design concerns thinking about what type of empirical material will help you understand a certain phenomenon.

on the basis of previous studies, questions are formulated which are then put to representatives of a representative selection of healthcare institutions.

You can relate to the empirical material you are gathering in different ways. Looking for patterns and causalities (predictive purpose) entails perceiving the empirical material (*explanans*) as a reflection of the empirical phenomenon (*explanandum*).

The purpose of the research design, then, must be to gather material which, as correctly and truly as possible, reflects the *entire* phenomenon being represented by the selection or example that you choose. This is how people reason when, in the example above, they choose to do a *survey study*. Then, they do a quantitative study of healthcare institutions which reflects, in a well-reasoned way, *the entire population*, i.e. all the healthcare institutions that exist.

> **Explanandum and explanans**
>
> *Explanandum*
> Something needing to be understood – the *phenomenon* you wish to say something about. Expressed in the purpose.
>
> *Explanans*
> What you use in order to understand – the *example(s)* (the empirical material) used to say something about the phenomenon. Expressed in the choice of method.

Trying to develop an understanding of a phenomenon by pointing to its complexity and ambiguity entails perceiving the empirical material as multifaceted, maybe even impossible to capture purely objectively. The purpose of the research design, in this case, must be to gather profound material in which as many aspects as possible are included. This is how people reason when, in the example above, they choose to do a *case study* in which they thoroughly gather empirics from a case (*explanans*) and use them to say something about the phenomenon (*explanandum*) (see the example below). The choice of this research design is based implicitly on an approach that entails considering it impossible – and maybe even uninteresting – to capture all the aspects of a phenomenon. Put differently: *explanans* can never fully reflect *explanandum*. Thus, this type of study does not claim, either, to tell the truth about the phenomenon (*explanandum*) you wish to study.

> **Choosing a research design**
>
> A group of researchers at Linköping University were interested in finding out whether engineering students showed the same degree of empathy as other students. A survey was formulated which actually showed that this was not the case. The answers given by the engineering students pointed to their actually having a lower degree of empathy than other groups of students.
>
> In the article where the researchers account for their findings, they point out that this does not necessarily mean that engineering students are really less empathic. Maybe it is their education that forms them in such a way that they become less empathic, reason the researchers. Exactly how things are in that regard cannot, however, be discerned from the data material that the survey study had generated.
>
> To develop a more profound understanding of how it is that engineering students, in a survey, demonstrate a lower degree of empathy than other students, we have to use another research design – one that focuses on understanding rather than measurability. One possible design would be the case study in which a few technologists are chosen for interview, individually or in focus groups, or observations are made, in order to find out how they reason and act.

SOURCE: RASOAL ET AL. (2012), EMPATHY AMONG STUDENTS IN ENGINEERING PROGRAMS, *EUROPEAN JOURNAL OF ENGINEERING EDUCATION*, 37 (5), PP 427–435.

When writing your degree project on the basis of an assignment given by a company, or some other organization external to the university, you may think that the case and the empirical material you should be gathering are given. If engineering company X wants you to find out whether or not it is wise to outsource its production to China, isn't it natural to study engineering company X? But this is not necessarily the case. It depends on how the problematization is formulated, and what you choose to focus on in your purpose formulation. If you choose to do a predictive study, in which the purpose is to describe the financial consequences of closing a factory, then you will probably use financial data from other, similar closures and juxtapose these with data from engineering company X. In that case, consequently, the empirics constituting the *explanans* will be taken from elsewhere.

In what follows, we present three of the most common types of research designs in the field of social science:

- the case study
- the quantitative study
- experiments and action research.

Case study

As described above, the case study is a type of research design. It is *not* about what type of empirical material you are gathering in. You can thus do case studies where the empirics consist of words as well as case studies where the empirics consist of numbers. When doing a case study, you can collect different types of data, in different ways.

What is case study methodology – what is a case?

The case study has several characteristics. Firstly, it entails choosing *one or more individual examples*, i.e. one or more cases (*explanans*), in order to say something about the phenomenon (*explanandum*). This means, secondly, that you are studying a phenomenon in *real life* (cf. the experiments below). Thirdly, the method incorporates gathering *sufficient amounts of information* on the example you have chosen so that you can use that example to investigate, explain, or describe the phenomenon that the thesis will be dealing with. Fourthly, you have to conduct the case study *systematically* as regards both choice of case and data gathering methods. You will also have to accurately account for, and justify, how you made your choices and how you conducted the case study in the method section of the paper.

The choice of case is extremely significant when it comes to the quality of the finished work, especially in respect of the consistency between purpose/question formulation and the analysis that you do. You either choose a case on the basis of what previous research and theories in the field say (depending on the question formulation, the choice might be between a typical and an untypical case) or "the case chooses the researcher", i.e. the case you have access to becomes the one you study.

> Being open to the possibility of a case study maybe leading to new discoveries is positive and part of the critical attitude that distinguishes social science work of high quality.

In both situations, but even more so in the latter, it is not unusual for the case you have chosen to study to turn out to contain something other than what you had initially envisaged. Then, you will have to be pragmatic and change your purpose/question formulation. This is not as unscientific as it sounds. Being open to the possibility of a case study maybe leading to new discoveries, based on which you realize that something other than what you initially envisaged studying is more interesting, merely shows sensitivity to the empirical material, which is positive and a part of the critical attitude that distinguishes social science work of high quality. Besides, it will in all probability lead to an improved, more interesting, and more relevant thesis for the client, too.

Why a case study?

Many theories in social science are based on case studies. The reason for this is the fact that the case study generates rich empirical material in which the complexity of reality is captured better than when, for example, experiments are conducted or a little bit of data is gathered about many cases, e.g. via a survey. The case study enables the discovery of new dimensions. Frequently, the case study is specifically used in inductive studies, i.e. studies where we allow theory to emerge from the empirical material. The result is often a theory which is new, or at least further developed, and empirically valid, since it has evolved from the analysis of the empirical data. The case study is suitable as a research method when we are open to discovering new dimensions – i.e. when the purpose is researching, explaining, or describing – and when we want to answer questions beginning with how or why.

The generalizability of the case study

How can empirical material from an individual case (*explanans*) be used to say something about a phenomenon (*explanandum*), which can look different from case to case? The criticism which is at times levelled at the case study as a research method is that it is subjective and primitive – that it is unscientific. And sure, there are examples of badly-conducted case studies, as there are of badly-conducted studies claiming to be scientific, and which have been conducted using another research design.

In order to counter this criticism, we need to be careful about how we choose cases and how we do our case study. Remember that a case study is characterized by *systematics* when it comes to choice of case, data gathering method, and analysis method, as well as a clear account of this in combination with both justifications of and reflections upon the choices made and how the research journey has gone.

Additionally, we need to reflect upon what we can say using our case study. A case study can never result in *statistical generalizability*. This is rather self-evident, really. On the basis of studying one (of several) cases, we cannot assert that the findings from this single case, with any statistical probability, will apply to all other cases, even if they are similar. On the other hand, on the basis of (well-conducted and described) case studies, we can speak of *analytical generalizability*. Analytical generalizability means that you discuss, in the analysis section (fittingly towards the end of the discussion section), the way in which the results of the case study may be applicable to other, similar cases. In order to do that in a good way, we need to account for our case in sufficient detail so that the reader can assess whether or not our line of reasoning holds water. Sometimes, we also use the word *extrapolation* to describe the kind of comparison that the reader, after having read both a well-developed description of a case (in the findings section of a paper) and the subsequent analysis (in the discussion section) will make on his/her own when evaluating how and to what extent the lessons drawn from the case in question are relevant to some other case – maybe a case that the reader is well aware of.

Types of case studies

In the methods literature, there are descriptions of many different kinds of case studies. In addition to the *exploratory* case study, the *comparative* case study is one of the most common types of case studies in degree project contexts. Here, several cases are studied (in degree project contexts, often two), and compared and contrasted. A *historical* case study can also be conducted in which there is

> "The type of case study chosen depends on the purpose/question formulation, but also on practical issues .

a focus on more general changes occurring over time in the field being investigated. One form of historical case study is the *longitudinal* one, entailing the following of a delimited case over a longer period of time in order to study changes. Longitudinal case studies are not, however, and for natural reasons, especially common in degree project contexts.

The type of case study chosen thus depends on the purpose/research

> **The good case study researcher**
>
> - is inquisitive
> - aspires to listen, observe, and get a feeling
> - is flexible
> - aspires to understand the questions and environment under study
> - is impartial.

question, but also on practical issues such as how much time is available, the kind of access to empirical material, and so on.

Conducting a case study – hard work and lots of data

One characteristic of the case study, as previously mentioned, is the gathering of lots of data about the case. This data gathering can take place in several ways: through interviews and observations, by collecting written documents in the form of reports, memos, records, personal documents and emails, by downloading web pages, taking photos and videos, and so on. The data gathering methods chosen depend on the type of data best believed to assist in answering the question formulations that have been framed.

Regardless of which data gathering methods are chosen, a lot of data will be generated when doing a case study. This is, of course, an opportunity since a large volume of data, systematically gathered and well-focused in relation to the purpose/research question, can entail the possibility of saying a lot about what is being studied. However, a large dataset also brings a challenge, especially for a degree project author with a narrow timeframe within which to complete his/her degree project. When doing a case study, it is thus of the utmost importance to get the study started early on during the degree project and to continuously reflect upon how the content of the material being obtained tallies with the purpose/research question in order that these can be adjusted and some thought can be given to whether or not the right data gathering methods are in use.

Quantitative studies

As indicated in the introduction to this section on research design, instead of studying one or just a few cases, we can do a quantitative study in which we collect quantitative data on a larger number of examples (*explanans*) of the phenomenon (*explanandum*) we want to say something about. This takes place either with the help of fresh data, known as primary data, e.g. using a survey in which the answers are converted into numbers, or with the help of already-gathered data, known as secondary data, where the data comes, for instance, from various registers, official statistics, or various publications.

For practical reasons, it might be appropriate to collaborate, as a degree project author, with someone else when doing a quantitative study since the volume of data being gathered can be large. It might also be appropriate to process the empirical material digitally using suitable software.

Why a quantitative study?

The advantage of quantitative studies is that they can provide a good overview of a phenomenon. The quantitative study can thus work well as an initial stage of an explorative study in order to answer questions starting with *who*, where there is a possibility during the next stage of entering into individual cases in detail in order to find out *how* or *why*. The quantitative study can also work as Stage 2 following an explorative inductive study in which, based on the first study, one or more hypotheses are put forward which are then tested using a deductive quantitative study. The quantitative study can thus be used both inductively and deductively and suits all four types of purposes we have talked about: i.e. exploring, explaining, describing, and predicting.

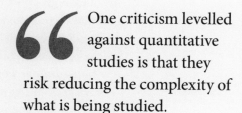

> One criticism levelled against quantitative studies is that they risk reducing the complexity of what is being studied.

One criticism levelled against quantitative studies is that they risk reducing the complexity of what is being studied. On the other hand, the quantitative study enables the measurement of strength in between various factors, which is often very valuable, especially when many factors are affecting a phenomenon. It is, of course, of great significance to really make sure that all factors are measured in the quantitative study. If this is not done, there will be a risk that some factor that has been omitted will affect the values which measure the connections between the factors actually being measured.

How is sampling conducted in a quantitative study?

When doing a quantitative study, the first thing to be defined is the *population*, i.e. all examples of the phenomenon to be studied. Since most of the time it is not possible to study the entire population, sampling will have to be done; thus, this research design is sometimes called the sample study or sample investigation. Sampling can be done in two ways: random and non-random.

Random sampling entails selecting from the entire population. Sampling can be done in an unsystematic way (known as *simple random sampling*), or there can be a certain level of systematic choice, e.g. by choosing every fifth, tenth, or fiftieth (or similar) company, organization, or individual from a given list (known as *systematic sampling*). A further alternative is investigating all examples in a certain group – if the population can be considered naturally divided into groups, i.e. *cluster samples*. Finally, we can divide up the population ourselves into groups, e.g. by age, gender, income, profession, and so on, where people are concerned, or into numbers of employees, turnover, age, industry or similar, where companies and organizations are concerned, subsequently randomly taking a sample from each group which is just as big as that group's share of the entire population (known as *stratified random sampling*). Which groups are created will depend on what is to be studied and should be indicated in the problematization and purpose.

Non-random sampling entails investigating the examples most easily accessible (known as *convenience sampling*) or those willing to participate. The advantage of non-random sampling is the relative ease with which empirical supportive data can be obtained for a study. A type of non-random sampling is *snowball sampling*, i.e. using respondents with whom contact was initially established in order to establish contact with additional respondents. This may be suitable when it is not possible to create an accessible sampling framework for the population. The problem with non-random sampling is that the conclusions reached may not necessarily represent any others than the companies, organizations, or people included in the study, which can be a criticism of this type of sampling procedure. Here, it is important to ensure that the problematization and purpose tally with what is actually being studied: otherwise there will be problems with the validity of the study. On the other hand, non-random sampling can be suitable when time is of the essence, or when a study is not primarily aimed at providing exactly correct answers.

When writing a degree project, sampling is most frequently primarily dependent on practical circumstances, e.g. how much time is available to gather material and how high the level of acceptance is, both in yourself and in your supervisor, with regard to the investigation being less exact. If you do not have so much time but you do have a high level of acceptance of the study possibly being to some extent misleading, you can do non-random sampling.

If you are writing your degree project in close collaboration with a company or another organization, and want that company or organization to be included in the sampling of companies or organizations, this means that you have done non-random sampling. However, you will still be able to randomly sample respondents, i.e. those taking part in the investigation, at the company or organization.

Just how big the sample should be cannot be stated exactly; it depends on the study. Generally speaking, it can be said that if the population is heterogeneous – if the examples of the phenomenon that is of interest are very different from one another – then a larger sample would be needed than if the population were homogeneous (if the examples were similar to one another). In the former case, the so-called *standard deviation* is large, which means that the average deviation of the individual examples from the population's average value is large, which entails an increased risk that the sample which is being taken will not be representative. Thus, it is reasonable to have a larger sample. In order to know whether the population is heterogeneous or homogeneous – how the standard deviation looks – we need to form an opinion about the population. In some cases, we can be inspired by how other studies have been conducted – what sampling others have done and why – and sometimes we have to gather information ourselves in order to be able to determine this.

The generalizability of the quantitative study

The strength of the quantitative study is that it is possible to discuss its generalizability purely statistically. However, this will depend on how the sampling was conducted. If we do non-random sampling, then the statistical generalizability will be low. If we do random sampling, then the probability

> The strength of the quantitative study is that it is possible to discuss its generalizability purely statistically.

will be greater that this sample is representative of the whole population, and thus the generalizability will be greater. But the size of the sample matters, too, as discussed above.

Conducting a quantitative study – solid reading and lots of numbers

One characteristic of the quantitative study is tht it requires putting a lot of effort and energy, prior to conducting the study itself, into making well-considered choices. This entails having to read up thoroughly on the population, as it was studied in previous research and as it would appear if you read up on it on the Internet, in journals, trade journals, or other forms of literature. It may also be appropriate to conduct interviews with those who are familiar with the population. The aim of all this is to choose a good and justified sample in relation to the question and purpose.

As a quantitative study entails gathering quantitative data, you also have to be prepared for the analysis work focusing on numbers. This work can entail either working with various descriptive techniques (e.g. average value, median, frequency) or working with various techniques of attempting to generalize on the basis of the sample that you have (e.g. techniques of showing correlations, i.e. links, between two or more dimensions of the material).

> The experiment is similar to a case study in terms of research but entails studying a created and in various ways controlled situation.

Experiments and action research

Experiments occur in social science research when the purpose is predictive – often when a hypothesis is to be tested (using a deductive arrangement). The experiment entails creating a situation (*explanans*) that is typical of the phenomenon to be commented on (*explanandum*) and studying what happens when the situation is manipulated in some way. The experiment is similar to a case study in terms of research design in that it also entails studying a case; however, while the case study methodology is used to study something in real life, the experiment entails studying a created and in various ways controlled situation.

One variant of the experiment, or perhaps instead a cross between an experiment and a case study, is *action research*, which is a method developed by the social psychologist Kurt Lewin in the 1940s. Action research entails studying a process that you yourself are involved in and which, additionally, you are attempting to change during your study. Action research includes:

1. identifying a problem
2. gathering data and making a preliminary analysis (diagnosis) of what is causing the problem
3. reconciling the diagnosis with the client
4. jointly arriving at suitable measures to rectify the problem
5. acting
6. gathering data about how the process has been going and making a new analysis
7. reconciling with the client.

> **Expectation management during change processes**
>
> The student, who had concluded that she was going to study expectation management during change processes in connection with the introduction of a new IT solution (see p. 44), was asked by the agency where she conducted her case study to propose a model for how they might be able to work with this. Based on the available literature, she developed a process model of expectation management which she tested on a small scale at the agency for a month or so by means of a manger working in accordance with the model under her supervision. Over the course of that month, she conducted observations and interviews with those involved. She then evaluated the results, made the necessary changes to the model, and presented it to the agency. In her thesis, she accounted for how she had created the initial model, how she had conducted the study, and what the observations and interviews had provided. In the analysis section, she discussed the way in which the investigation had led to changes to the model, which she also presented in the text.

Action research is always about real life problems and aims to assist in their solution. It is an iterative process whereby a contribution is made in the form of changed practice – and in some cases developed theory. The research does not conclude until after Stage 7 (see above), entailing that action research is not always possible as a research design for a degree project author, who often has a finite amount of time.

CHOICE OF DATA GATHERING METHOD

Over and above the choice of research design, you also need to choose a way of collecting data. In the following section, you can acquaint yourself with five of the commonest data gathering methodologies for students writing degree projects with a social science slant:

- document gathering methodology
- interview methodology
- focus group methodology
- observation methodology
- survey methodology.

The empirical material you collect – your sources – must be critically scrutinized, regardless of data gathering method, using the criteria we present in the section "Source evaluation and criticism" on pp. 124–126.

Document gathering

You can, of course, make use of different kinds of documents as empirical material – both as primary and secondary sources. Examples of documents include letters/emails, autobiographies, photos, official documents, documents from organizations, virtual products and mass media products. What documents you gather will depend on your purpose and question formulation. When doing a case study, it is common to gather documents as one of many data gathering methods. But you can also work solely with documents as empirical material.

> Gathering empirics via interviews is one of the most common methods used in qualitative social science research.

Interview methodology

Gathering empirics via interviews is one of the most common methods used in qualitative social science research, not just in degree project contexts. The reason for this is that it is possible, using relatively simple means, to learn more about how individuals reason in different research questions.. If the interview is of an open nature, there will also be the opportunity to obtain new ideas regarding dimensions of the phenomenon under study, which can bring about a desire to pose new and different questions. The interview provides good opportunities to make unexpected discoveries, which is an important dimension of qualitative research. Conducting interviews at the start of the degree project process can be a good idea because it can help refine the problem that the thesis will be studying.

When is interview methodology suitable?

The interview as a research method is suitable when there is an interest in developing a deeper understanding of a phenomenon, when there is a desire to discover new dimensions of what is being studied, and when there is an interest in multiplicity.

> **Diversity management at knowledge-intensive companies**
>
> A student had decided, on the basis of self-interest, to study how knowledge-intensive companies work with diversity management, i.e. to meet and benefit from the ethnic, cultural, age-related, and gender-related multiplicity of these companies. She conducted eight semi-structured interviews with representatives of two high-tech companies, which gave her the opportunity to follow up, on the basis of the question areas she had formulated in her interview guide, the informants' answers and to deepen her understanding of how they worked.

Different types of interviews

Note that we are using here the word *interview* synonymously with qualitative interview. Sometimes, there is also talk of the quantitative interview, also called the structured interview, which is when the interviewer fills in a survey together with his/her informants. In this case, there is a preprinted form – a questionnaire – containing yes and no questions and/or questions in which the informants are asked to grade their answers on a scale, and in which the interviewer fills in the answers together with the informants. Thus, the structured quantitative interview is a way of distributing a survey, and also of gaining the possibility of posing follow-up questions (which will then be semi-structured in nature). If this is done by phone, it is called a *phone survey*. However, the purpose is, as with all survey studies, primarily to create empirical data in the form of measurable quantities – numbers

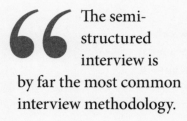

> The semi-structured interview is by far the most common interview methodology.

– as opposed to the qualitative interview which is aimed at creating qualitative data and thus focuses on words.

A qualitative interview can be either unstructured or semi-structured. In an *unstructured interview,* it is not decided in advance what is to be found out – there is only an overarching topic for the interview. This type of very open interview is suitable in the beginning of an empirical study when there is an unbiased desire, and need, to explore a subject field as described above.

It is, however, the semi-structured interview that is by far the most common when empirical material is being amassed using the interview methodology. A semi-structured interview is organized around a number of themes or question areas, determined in advance and frequently written down in an *interview guide*. The interview guide must not be too detailed or too extensive; it often fits on a single slip of paper. The various question areas are then dealt with in the order that best presents itself during the course of the interview. This means that, during the interview, your aspiration is flexibility vis-à-vis the informant, raising the questions in the order that feels most natural in relation to how the informant answers.

Most of the questions posed during a semi-structured interview, as well as during an unstructured one, are not formulated beforehand but created during the course of the interview. Depending on the interplay between you and your informant, this can be easier or more difficult. Some informants take up a lot of space during conversations, talking gladly and plentifully. Others have to have the answers extracted from them by means of asking lots of questions.

The questions you ask can be of different kinds. They can, for instance, be *introductory*, *probing*, *interpreting*, or *specifying*. You can also use *silence* as a means of pushing the interview forwards.

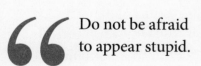

> Do not be afraid to appear stupid.

Being silent provides the informant with the time to reflect; it is frequently during pauses that the most interesting statements emerge. Exactly as in other kinds of conversations, you can also push the interview forward by following up on what the informant says by nodding in agreement, saying "mmm", repeating what the informant says, and so on. The interview guide, like your behaviour during the interview, must stimulate good interplay between you and your informant. It is, however, very important that you remain inquisitively critical during the interview and that you do not let anything you do not understand slip past you. Do not be afraid to appear stupid. Even though you should, of course, be as read up as possible ahead of an interview, you should not be afraid to ask questions. It is only what the informant says that counts in the empirical material, not what you are thinking.

> **Interview questions of different kinds**
>
> *Introductory*
> Can you tell me about …? Can you describe …?
>
> *Probing*
> Can you give me an example? Can you describe what you mean?
>
> *Interpreting*
> Do you mean that …? Could we say that …?
>
> *Specifying*
> What did you do then? What did you think then? How did you react then?

Conducting an interview has been likened to sitting and conversing with someone on a stage. Everything you say and do has an impact on how the interview evolves. It is best to sit in a secluded place, to have mobile phones in silent mode (it is far from cheeky to ask the informant to turn his/her mobile off during the interview). It is a matter of creating the best possible prerequisites for the interview.

Extent of the interviews

A question that degree project authors often ask themselves is how many interviews have to be conducted within the framework of a degree project. Some universities have guidelines for this; we have heard a figure of 10–15 interviews for a degree project on the Master's level. But this is just a pointer. Exactly how many interviews you will need to conduct depends on their quality. If you have informants who are completely "right" in relation to your problematization, and if these give long and, vis-à-vis your research question, content-rich answers, then maybe you will achieve empirical

> " If you have informants who are completely "right", then maybe you will achieve empirical saturation with fewer interviews.

saturation, as it is called, with fewer interviews. Our experience is that committed students often notice when they have achieved this saturation – when the interviews no longer provide any new or relevant information. Be careful, however, that you do not draw this conclusion purely on the basis of having conducted sloppy interviews.

Documenting the interviews

Documenting the interviews is of the utmost important when it comes to the possibility of returning to what was said. The most usual thing today is to use a mobile phone to record conversations, but there are also other technologies for this purpose. Purely making a recording, however, is not to be recommended; it is also a good thing to take notes in parallel with that. If there are two of you, it may be beneficial for one of you to focus on the interviewing itself while the other takes notes. Using these notes, you can easily return, during the course of the interview, and follow up on things that the informant said.

> Having a positive attitude matters when it comes to conducting a good interview.

Quality criteria for interview studies – scientific legitimacy

As indicated above, the qualitative interview is a social encounter, and its quality is thus dependent on the extent to which you, as the interviewer, are able to create a pleasant and open conversation. Having a positive attitude, looking the informant in the eye, saying hello and goodbye politely, being able to pose and follow up questions in order to obtain deep and rich answers – all of this matters when it comes to conducting a good interview, and thus creating the prerequisites for an interview that is of high quality. The simple quality criterion for a good interview is that it contains brief questions and long, exhaustive answers that are relevant to the research question.

When regarded as empirical material, you can view the interview material in two ways. You either understand it to be the truth, whereby you, to use Steinar Kvale's formulation, are a "prospector" looking for a certain mineral, i.e. the fragments of the interview on the basis of which you are able to build the answer to your research question, or you regard the interview material as the documentation of a journey that you, together with the informant, have made through the terrain constituted by the phenomenon you are interested in studying.

In the former case, you, as the interviewer, are seeking *comparable answers* between different informants; you seek arithmetical reliability between different interviews whereby you measure the congruity between them (which is noticeable during the Produce Phase in formulations such as: "The majority of informants are of the opinion that …", "Ten out of twelve of those interviewed are of the opinion that …").

In the latter case, you are looking for *multiplicity* and *complexity*, which will also be reflected in the findings section of your paper where you account for your empirics. This view entails that, rather than aspiring towards objectivity, in the sense that you are entirely removed from what you are studying, you aspire towards being impartial; i.e. that you remain – even though you are pleasant and polite in your dealings with the informants – inquisitively critical throughout the entire process.

> **Worth bearing in mind before conducting an interview**
>
> - Practice posing open and follow-up questions.
> - Confirm the agreed time and place via email.
> - Read up on those you are going to meet.
> - Prepare yourself so that you know how to present yourself and the aim of the interview so that you may quickly be able to get the interview started and your interviewee talking.
> - Come to the interview well rested – the quality of the interview will be better then.
> - Reflect upon the ethical dimensions of your interview and prepare to answer questions about confidentiality.
> - Decide how you will be documenting your interviews and make sure that you have tools that are in working order and correct (pens, writing pads, camera, recording equipment, and so on).

Focus group methodology

One variant of the interview is the focus group interview. This is an interview conducted with a group of several people. The focus group interview is suitable when you want general information about something, when you want to test an idea, or when you want to provide the stimulus for new ideas.

Before conducting a focus group interview, you will have to think about who the members of the group will be, and how big it will be. A *heterogeneous group* brings a greater possibility of developing new ideas but can also place major demands on whoever is moderating the discussion, who will have to ensure that everyone gets to speak and that the conversation does not become superficial because people are careful about expressing their thoughts in a group where people view the subject that is up for discussion differently. A *homogeneous group* brings greater possibilities of delving more deeply into a certain question, but it also entails the risk that the conversation may be restricted content-wise due to the participants having similar opinions. Whatever is best here will depend on the aim of the focus group interview in relation to the purpose and question formulation you have set for your degree project.

The *size of the group* is another thing to reflect upon ahead of the interview. The suitable number of people will depend on whether or not the participants know each other, whether the group is heterogeneous or homogeneous, and whether those invited are talkative or quiet. You should reflect, in the method section of your paper, upon these dimensions of the focus group interview and what part they have played in your findings.

> **Use of an app for energy consumption**
>
> In order to find out how people used a mobile app to keep track of their energy consumption at home, a focus group survey was conducted using six groups with 5–7 participants in each. Half of the groups consisted of people who used the app while the other half consisted of people who did not. Those using the app were asked to say how they used it, with a discussion about that ensuing. Those not using the app first had to say whether or not they had used other similar apps; then, the researcher led the conversation on to mobile apps for energy consumption and other ways of keeping tabs on one's energy consumption. Each conversation, lasting between 90 and 120 minutes, led to an enhanced understanding of people's relationships with mobile apps, which was the phenomenon that the study was dealing with.

Conducting a focus group interview involves major challenges since the conversation has to be moderated by means of intermediating between the participants while simultaneously focusing on the conversation so that it is about what you want to discuss (what you have noted in your interview guide). This entails having to focus on what is being said so that you can follow up on what the participants are saying, in the same manner as during a one-person interview. Recording the focus group conversation is definitively to be recommended. Just make sure, however, that you clearly note down which voice belongs to whom, if your recording is audio only, so that you are aware of this when listening to the recording later on. Video filming the focus group interview is one alternative and, even though many think it feels slightly unpleasant to be filmed, people usually forget the video camera after a while, especially if it is on a tripod in a corner of the room, which is to be recommended.

> Conducting a focus group interview involves major challenges.

Worth bearing in mind before conducting a focus group interview

- Carefully plan the composition of the group (which participants and how many) on the basis of your purpose and question formulation. This pre-requires knowing, or at least knowing about, the participants.
- Think through how you are going to behave during the interview itself – the extent to which you will be active and/or passive. This is dependent on what you want to achieve with the focus group interview, but also on the composition of the group.

Observation methodology

Observation methodology involves systematically observing and documenting, over a long period of time, what is happening at a company or within some other organization. This method has its roots in ethnology, i.e. the scientific discipline that studies foreign cultures. The first group of better-known studies conducted by organizations using observation as a method included the Hawthorne studies, which were conducted at the Western Electric factory in Hawthorne outside Chicago during the 1920s and 30s. Their purpose was to find out how to increase production at that factory and, following several studies, it was established that the involvement of management in the workforce played a major part in its productivity. At the same time, it was seen that there was a boundary vis-à-vis how much the formal leader was

> **Observation methodology involves systematically observing and documenting, over a long period of time, what is happening.**

able to do; there were also informal leaders among the workers who played a major part in achieving the maximum level of worker performance. This study was crucial in highlighting issues of leadership, group dynamics, and staff welfare, leading to, for instance, the development of the theoretical and practical field nowadays called HR – human resources.

From the 1940s to the 1980s, observation methodology was not particularly common in the study of organizations. However, with the publication of Gideon Kunda's book, *Engineering Culture*, on a study of the organizational culture of an Israeli IT company in 1992, interest was aroused once again. In step with the growing interest in the significance of culture in understanding what happens at organizations and other companies, and for symbolic and interpretive perspectives, observation as a method for understanding organizations has become more and more common since the 1990s.

One variant of ethnography is *netnography*. This studies digital communities, i.e. different groups of people and their online interactions on the web, providing some interesting opportunities for gaining insight into and analysing that type of context.

When is observation methodology suitable?

Observation as a method is suitable when the questions you want answered are of an exploratory nature, e.g. questions about how and what people do and what appearance day-to-day work and management have at an organization. During observation, you observe and document what is happening; behaviours and actions of people and groups, but also what people say, how they say it, and the physical environment.

> **Observation as a method is suitable when the questions you want answered are of an exploratory nature.**

> **How are project models used in practice?**
>
> In order to find out how project models are used in day-to-day work, one student decided to conduct a study using observation methodology at the consultancy where she had also recently started working part time in parallel with completing her university studies. For three weeks, she shadowed a project manager, noting down everything he did in a field diary which she carried around with her. In the evenings, she fleshed out her notes and wrote reflectively about what she had seen during the days. She also collected documents about the projects that the project manager was involved in, as well as documents showing the company's project model. Her analysis of the material showed that the project model fulfilled several functions of the project manager's day-to-day work as a tool for planning tool, governance, and communications.

Different roles as an observer

You can have different roles as an observer depending on the extent to which you interact with those you are studying, as well as the extent to which you tell them why you are there. The two commonest observer roles in research today are *participant as observer* (also called participating observer), i.e. working at a place and simultaneously informing that you are observing, and *observer as participant,* which entails observing and interacting with those you are studying by asking questions, chatting, and so on, but not by doing any work in parallel with your observations.

In both roles, you can conduct *ethnographic interviews*, which are interviews arising spontaneously during the observation situations, e.g. when you are accompanying your informants to/from meetings, having lunch or coffee breaks, and so on.

The traditional observation method entails studying groups of people in a certain place. But you can also *shadow* a person moving between different environments, or follow an object, e.g. an artefact or a project, rather than a person. Following an individual is demanding, not least because it is very tiring to document what you observe. Following an object is also difficult since it can spread out – especially if it is not a physical object but a project – into many different places simultaneously.

Which role you take on as an observer will depend on what you want

> The purpose of the observation is to obtain a view of things, and then you will need to adopt an inquisitive and open attitude.

to study and which questions you are seeking answers to. Regardless of what type of role you take on, however, you should aspire to *alienate* what you are studying. That means aspiring to perceive what you are observing as alien, as different. The purpose of the observation is, of course, to obtain a view of things, and then you will need to adopt an inquisitive and open attitude by which means what happens is regarded as new and alien. This can be a challenge if you are observing a company or another organization which, right from the very start of the observation, you are very familiar with. Then, you may need to work on "alienating" the day-to-day stuff; you will need to consciously try to regard what happens as alien even though it is not.

Extent of the observation

The extent of the observation depends, of course, on what your research question is and how much empirical material you need. If you are writing your paper on the basis of an assignment given by a company, or another organization external to the university, and you have decided to conduct the study there, then you will often have good access to the company/organization. Otherwise, the issue of access can be tricky if you want to do an observation, not least for reasons of confidentiality. Many companies, as well as other organizations for that matter, are often dubious about someone observing them, and many individuals feel that it is unpleasant to be observed. These are dimensions you will have to consider, investigate, and take a stance on before deciding to conduct an observation study.

It is difficult to clearly and quantitatively gauge how long an observation study has to go on for in order for it to provide enough material. As an observer, you will have to evaluate the material you have – best done by continuously reflecting upon the content of the material in relation to your question formulation – in order to be able to determine when you have enough empirics. As a degree project author, you will also, of course, have a timeframe set by your supervisor and/or seminar leader to keep to, which also gives an indication of when the observation study has to conclude in order for you to have time to complete the Produce Phase of your degree project writing – a phase that requires a lot of time and brain racking.

Documenting your observations

When observing, you have to document your observations carefully. This is very important as it is the documentation that will be used later on as supportive data during analysis. It is commonplace to make *field notes*, i.e. to take notes while work is going on in a fit-for-purpose exercise book or "diary". Naturally, this can also be done digitally on a computer, tablet, or mobile phone. The important thing is using a technique by which you can write a lot quickly. It would also be good if you could document using a *camera* and possibly a *video camera*.

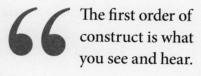

> The first order of construct is what you see and hear.

When documenting your observations, it is very important to differentiate between what is called the *first and second order of construct*. The first order of construct is what you see and hear – the pure documentation of what happens. Sometimes, this is called *operational data*. The second order of construct is your own reflections concerning what you see and hear.

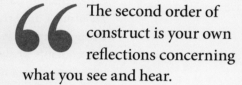

> The second order of construct is your own reflections concerning what you see and hear.

Examples of the first and second order of construct

First order
It's sunny outside (observation that can be verified by others)

Second order
What nice weather we're having today! (subjective interpretation that many, but not all, might agree with)

The term *construct* is used to emphasize that what is being documented when an observation is being made is not identical with reality; the first order of construct, too, is a construction of reality mediated via someone's observations. The first order of construct must, however, be something that those who were present can agree on and verify. The second order is the observer's own reflections on – interpretations of – what has occurred.

When making notes by hand in an exercise book, you may find it beneficial to use different coloured pens to mark the difference between the first and second order of construct. You might also make a habit of writing the first order of construct (the observations) on the left of facing pages and the second order (your own reflections) on the right. The most important thing, of course, is to focus on writing the first order of construct when observing; you can always write the second order later on. But, if you have time, it would also be good to make a note of these as they can sometimes constitute the embryo of an analysis, which may benefit you later on when analysing your material.

Quality criteria for observation studies – scientific legitimacy

Conducting an observation is very demanding since you also have to document, in parallel with your observations, what you are experiencing. Carrying out the documentation systematically is of the utmost importance as this forms part of the scientific legitimacy. Being systematic as regards documentation entails being able, when subsequently analysing and writing up your study, to report on your observations in a manner that exudes *authenticity* and *credibility*.

Authenticity is achieved through being able to reproduce *thick descriptions* of what has been observed. The text you write has to be sufficiently detail-rich, credible, and genuine in order for the reader to be convinced that you have actually been there making an observation. At the same time, you naturally cannot recap all the details, which is something of a challenge, of course, because you have to make a choice regarding which details you want to include and which you do not.

Over and above authenticity and credibility, *criticality* is usually counted as a quality criterion when scrutinizing observation studies. The question you should be asking yourself here is: Does your text work (as written on the basis of observations) in such a way that it explores and calls into question taken-for-granted ideas and views? Criticality vis-à-vis the empirical material is the basis of all scientific activities; however, as regards observations, it is perhaps of extra importance since the researcher could otherwise be accused of being partial or biased. This problem is known as *going native* – i.e. that the student becomes one of those being studied.

Becoming one of those you are studying need not be a problem – if you retain your independence as a researcher. Some are of the opinion that going native can even be advantageous as you gain better access to the environment and an improved understanding of how those you are studying think. Independence, expressed via criticality (impartial and inquisitive questioning) is necessary, however, in order for an observation study to be deemed scientific.

If you feel that scientific legitimacy is achieved via objectivity, then observation, as a scientific method, can be called into question, however. By definition, observation is subjective; it is the observer's subjective perspective that forms the point of departure for the analysis. Advocates of observation as a method often assert, however, that subjectivity in research, especially social science research, is impossible while some are of the opinion that it is not even desirable. Rather, we should affirm the subjectivity and instead reflect upon what it contributes to research. The purpose of observation is not, it is felt, to tell the truth about what is being studied but to provide a developed and well-founded picture of the phenomenon using what has been experienced, i.e. what has been observed.

> By definition, observation is subjective; it is the observer's subjective perspective that forms the point of departure for the analysis.

Worth bearing in mind before conducting an observation

- Practice differentiating between the two orders of construct.
- Practice "alienating" what you are studying.
- Make sure you are well-rested when observing – it is very tiring to conduct observations.
- Make preparations in order to know how to present yourself, what you are observing, and why, in case anyone asks so that you may quickly be able to say this and then focus on observing.
- Reflect upon the ethical dimensions of your observation (see the section on ethics on p. 37).
- Decide how you will document your observations and ensure that you have correct and working tools (pens, writing blocks, camera, recording equipment, and so on).

Survey methodology

Survey methodology entails asking a number of people (known as respondents) to answer a number of questions on a form, known as a questionnaire. The questions are based on your research question and on the hypothesis or hypotheses you may have formulated, on the basis of previous research. Most often, the survey questions have several predetermined alternative answers, but there can also be open-ended questions to which the respondents themselves specify their

> Survey methodology is suitable when you want to do a quantitative study, when you are interested in finding a general answer to the question formulation, hypothesis, or model you have formulated, or when you are interested in gaining an overview of all the examples that shed light on the phenomenon.

own answers. Once you have collected the answers, you compile them, most frequently using some form of software, by giving each alternative answer a numerical value. Using the software, you will then be able to set the various numerical values in relation to each other and thus understand the relationship between the different questions you have asked in order to answer your question formulation in that manner.

When is survey methodology suitable?

Survey methodology is suitable when you want to do a quantitative study, when you are interested in finding a general answer to the question formulation, hypothesis, or model you have formulated, or when you are interested in gaining an overview of all the examples (*explanans*) that shed light on the phenomenon (*explanandum*).

Using a questionnaire, you can thus obtain a large volume of quantitative data on the phenomenon you are interested in. This is both a strength and a weakness, something which you should be aware of when choosing this method as it takes time to compile and analyse the empirical material. You should also make sure that you have knowledge of as well as access to good software for creating, and possibly also distributing and compiling, the questionnaire.

Designing a questionnaire – operationalizing

When designing a questionnaire, there are two things you should bear in mind: the formulation of the questions and the alternative answers as well as the questionnaire's overall design.

Exactly which questions a questionnaire is to include depends on the research question(s) that the study aims at answering. If you are using a deductive research approach, you may have additionally formulated one or more hypotheses on the basis of previous research, or you may have designed some kind of model which, for instance, shows connections, e.g. of the type cause and effect, between various factors. The problematization, hypothesis, or model includes different words (concepts) and it is both these and the relationship between them that the questions you formulate in the questionnaire are aimed at studying.

The activity of translating the terms in the question formulation, hypothesis, or model into questionnaire questions is called operationalizing. When working on the operationalization of your concepts, it is an advantage to allow yourself to be inspired by other previous studies. Regardless of whether you borrow questions from previous studies or create your own, operationalizing the words and concepts in the research questions or hypothesis is something that requires great consideration. The first step is to read up on already-existing theory. When you read up on the field, you will realise that there are many different kinds of theories, and that these emphasize different things. Here, you will have to choose the theory that seems most suited to the purpose of your study, and be able to justify why you are making that choice. Depending on which theory you have chosen, a certain term will be defined differently, and it is this definition that you will start out from when operationalizing the term.

Operationalizing leads to measurements or indicators of the terms we are interested in studying. The measurements or indicators of the terms are called variables. It is the variables that function as a point of departure for the statistical analysis. There are different types of variables, something which is connected with the fact that different variables pre-require different measuring scales. If the question can, for instance, be answered using two or more alternatives which cannot be ranked in any way, then the variable is *nominal*. If there are more than two answer alternatives that can be ranked, but have unequally large distances between them, then the variable is *ordinal*, but if there are more than two answer alternatives that can be ranked with equally large distances between them, then the variable is called a *ratio* or *interval variable*. (See Figure 11.)

Depending on which type of variable you have, you can generate different kinds of data and, in doing so, carry out different kinds of analyses. It is important to have several variables for each term in the hypothesis or model in order to make sure that they jointly capture all the dimensions of the concepts.

> The activity of translating the terms in the question formulation, hypothesis, or model into questionnaire questions is called operationalizing.

> Operationalizing leads to measurements or indicators of the terms you are interested in studying. The measurements or indicators of the terms are called variables.

Operationalizing innovativeness

A couple of students wanted to investigate the extent to which companies from various industries differ as regards innovativeness. In which industry are companies the most innovative? They decided to send out a questionnaire to a number of companies. In order to be able to formulate suitable questionnaire questions, they read up on different theories about innovation. In the OECD's Oslo Manual from 1997, innovation is defined as a new product, process, or form of organization. Operationalizing the term innovativeness in accordance with this theory would thus involve them asking questions about how many new products, services, and organizational processes the companies produce over a certain period of time.

The problem with this operationalization, they came to realise, is that companies in industries where products or services have short lifecycles (e.g. the ICT industry) appear very innovative compared to companies in industries where product/service lifecycles are a lot longer, or where it takes a lot longer to develop a product (e.g. pharmaceuticals or forestry).

Another theory about innovation which they read up on said that a company's innovativeness is linked to its position within the innovation system, i.e. the network of actors that collaborate in different ways in order for new products, services, and processes to emerge. Operationalizing the term innovativeness in accordance with this theory would thus entail the students formulating questions concerning the company's collaborations with other actors within what is defined as the innovation system. For instance, the students would then ask questions about how many (and possibly which) trade associations the company participates in (and possibly how), to what extent and in which ways the company is in contact with universities, if there is participation within some form of triple-helix collaboration, and so on.

The problem with this operationalization was that companies in certain industries have a lengthy history of collaborating with universities, while other industries, e.g. the service sector, do not, making the service sector appear less innovative.

Once the students had read up on theory, they realised that the purpose they had initially envisaged was way too comprehensive; they thus decided to study the degree of innovation within one and the same industry and to instead introduce size as a variable. In the final version of their thesis, their purpose formulation was: "The purpose of this thesis is to investigate whether or not the size of service companies (measured in terms of numbers of employees) correlates with these companies' degree of innovation (measured in terms of numbers of new services)."

When designing your questionnaire, you may benefit from using questions from previous studies. The questions will have already been tested and you know that they work. It may also be of interest to compare your results with those from other studies. Additionally, it may be an advantage if you obtain unexpected results since you will be able to go back and investigate how the variables, which in your own study are turning out differently, have behaved in other studies.

1. How many employees work at your company?
    _____ employees   NUMBER OF EMPLOYEES = RATIO/INTERVAL VARIABLE

2. Which of the following alternative best describes your operations? (mark one alternative)
    [ ] Wholesale and retail
    [ ] Real estate company and property management
    [ ] Information and communication
    [ ] Transportation and storage
    [ ] Legal, financial, architectural, or technical consultancy and R&D
    [ ] Financial and insurance
    [ ] Education, care, and social services
    [ ] Leasing, job agency, travel services, other support services
    [ ] Personal and cultural services
    [ ] Hotel and restaurant
    [ ] Advertising, market research, other operations in law, finances, science _____
    [ ] *Other (please specify)* _____   TYPE OF SERVICE COMPANY = NOMINAL VARIABLE

3. How often do you introduce a new service?
    [ ] Monthly
    [ ] Quarterly
    [ ] Biannually
    [ ] Annually
                    FREQUENCY OF INTRODUCTION OF NEW SERVICES = ORDINAL VARIABLE

4. Approximately how many man-hours per week do you dedicate to developing new services?
    _____ hours/week
                    MAN-HOURS DEDICATED TO SERVICE DEVELOPMENT = RATIO/INTERVAL VARIABLE

FIGURE 11 Examples of questionnaire questions giving different types of variables.

> **MECE – mutually exclusive, collectively exhaustive**
>
> MECE is a thought tool for operationalizing terms which can be used to ensure that the questions included in a questionnaire really cover the concepts. *Mutually exclusive* means that the questions exclude one another, i.e. they must not be coherent with several terms. *Collectively exhaustive* means that they must be jointly extensive, i.e. they must jointly capture all the terms in the research question, hypothesis, or model.

It is important when formulating questionnaire questions that the questions are not perceived to be value laden and that they do not contain negations, the latter being because the questions can then be misunderstood. It is also important that the questions are not transparent, i.e. that the respondent is unable to work out the answer that is being sought – which research question is being investigated.

It is also good to include some introductory questions with whose help you can make sure that the respondent really belongs to the population. If the respondent does not belong to the population, you will have to discard his/her answers. When discussing the loss of answers later on, it will be good to be able to compare the distribution of answers on the basis of gender, age, and so on vis-à-vis the population.

> *How you formulate the questions and the different alternative answers is of the utmost importance when it comes to producing a good questionnaire.*

In order to make sure that the questionnaire is good, it would be wise to test it prior to sending it out to the sample group. They very best thing is to use the same type of respondent for the test questionnaire as for the live version since the feedback you get then will be the most relevant. It may also be wise to analyse the answers you obtain for tests so that you can see that the questions are evenly distributed for each variable.

How you formulate the questions and the different alternative answers is of the utmost importance when it comes to producing a good questionnaire. In addition, the questionnaire's overall design should make it easy for the respondents to answer the questions. Moreover, it is also important that the questionnaire looks business-like and credible. Find out if you can use your university logo and make sure to enclose a cover letter in which you briefly describe the purpose of the questionnaire, and how you plan to use the results.

> **A good questionnaire is characterized by**
>
> - Carefully thought-out and relevant (in relation to the purpose of the thesis) questions (not more questions on one issue; simple formulations; avoiding negations, value-laden and awkward words)
> - A brief but informative and clear description of what the questionnaire is aimed at achieving and how the answers will be used (see the section on ethics on p. 37)
> - Not too many questions
> - Numbered questions
> - A well thought-out order to the questions, possibly arranged in different sections with separate headings
> - Clear alternative answers
> - A visual layout which clarifies matters (e.g. the same distance between different alternative answers; a slightly greater distance to the next question).

Distribution

Today, it is common not just to design but also to distribute questionnaires digitally, either through channels like email or social media, or via special software. There are several freeware programs on

> **Distribution of a questionnaire on energy training and behaviour**
>
> On the basis of an assignment given by an environmental organization, a student decided to find out if an energy course brought any actual changes to the individuals taking part in it as regards energy usage at home. On the basis of this, additionally, improvement proposals were to be made regarding the content of the course offered by the environmental organization. A questionnaire was designed on the basis of existing literature in the field. When this was subsequently to be distributed digitally to a representative selection of those who had attended the course at the environmental organization, the student encountered problems. A large proportion of the participants consisted of elderly people who were either unable or unwilling to answer the questionnaire digitally. Instead, the student had to call the respondents and jointly fill in the questionnaire with them by phone.

the Internet with which you can design and distribute questionnaires, and analyse data in a simple way. This freeware, however, has certain limitations, e.g. with regard to the possible number of questions to ask, and security, which you should be aware of before making a choice. Make sure that the tool you use has functions allowing you to export the data generated by the questionnaire so that you can choose to run it in Excel or SPSS if you wish.

Choice of respondents, response frequency, and dropout analysis

As described above, it is important to think about choice when selecting who is to answer the questionnaire that will generate the primary data for a quantitative study. You also have to ensure that you send out the questionnaire to a sufficient number of respondents to enable an adequate number of responses to be received when a number of respondents choose not to respond. However, if you receive sporadic responses, you need not worry. The important thing is that the dropout is random, which you will be able to investigate if, for example, you have information about the respondents' age and gender. You can also follow up the dropout by calling some of those not responding and asking them to tell you why they have not responded.

The number of responses you obtain in relation to the number of questionnaires sent out constitutes the *response frequency*. A high response frequency is good, in contrast to a low one, as the empirical material in that case will be less representative of the entire population.

> Active response frequency = total number of responses/ total number of questionnaires sent out

When planning your degree project, you should attempt to counteract a low response frequency, e.g. by planning to send out a reminder after a certain amount of time to the respondents who have not responded prior to the deadline you have set. If you have the possibility, it may also be a good idea to increase the respondents' motivation to respond to the questionnaire by providing them with some form of reward, e.g. that they are given something for participating. Sometimes, it will be enough to clarify for the respondents that their participation will make a difference, e.g. that they will have the opportunity, by taking part, to influence a certain situation or issue. This must, of course, tie in with reality. You can also promise the respondents that they will be able to see the results once the investigation is finished, which is also a promise that has to be kept.

If the response frequency is less than 100 % (which is most often the case), you should do a *dropout analysis*. This dropout analysis entails thinking about why not all of the respondents answered the questionnaire and what you did in order to obtain as many responses as possible. Factors that can play a part here include the respondents not wanting to take part in the study, not reaching them with the questionnaire, or receiving illegible or incomplete answers, which can occur if the respondents fill in the questionnaire by hand or if the questions are difficult to understand. In the method section of the final version of the paper, the response frequency is to be accounted for. Here, the dropout analysis is also to be described and discussed in running text as a line of reasoning. In the concluding section of the thesis, the response frequency should also be juxtaposed with the degree to which the conclusions drawn can be deemed to be of general interest.

Telephone survey

If you only have a few questions, you can do a telephone survey. The advantage of a telephone survey is that it yields a high response frequency and that the risk of misunderstandings is lower as you will be able to clarify your questions if necessary. If several of you are collaborating on a degree project, it will be important, however, to make sure that you clarify the questions in the same way so that the answers do not vary due to the way in which those doing the telephone survey have formulated their clarifications.

A telephone survey can also function as a good test of a questionnaire before it is sent out. Being in direct contact with your respondents by phone, or in some other way, is always good as it increases the response frequency.

Quality criteria regarding questionnaire studies – scientific legitimacy

In contrast to the methods generating qualitative data (e.g. interview and observation), the questionnaire methodology often aims at developing statistically correct knowledge of the phenomenon you are interested in –or at least knowledge that you can assume to be generalizable to a larger part of the population than the selection you have covered through the respondents who have answered the questionnaire. In order for the questionnaire study to be deemed to have a high level of validity, the questionnaire questions must firstly show clear coherence with the question formulation/hypothesis. These connections are usually accounted for in the final version of the thesis; either in the method section or in an appendix and preferably in a table indicating which questionnaire questions are coherent with which parts of the research question(s) /hypothesis. A well-formulated and thought-out line of reasoning on both the response frequency and any dropout, as well as on the study's truth claims in respect of conclusions, will also enhance the validity since this affects the legitimacy of the knowledge.

In contrast to methods generating qualitative empirics, the quality of the questionnaire study can also be evaluated on the basis of its possibilities of generating the same results on another occasion. The *reproducibility* of the questionnaire study is, thus, another quality criterion linked to the reliability of the study. This quality criterion has to do with three dimensions. Firstly, the findings of the study must not be dependent on who has conducted it (known as *inter-rater reliability*). All the choices made along the way, e.g. as regards which theories have formed the foundation of the operationalization of terms to variables, must be accounted for and justified in a careful and clear manner. Secondly, the findings of the study must not be dependent on the time (known as *stability*). The questionnaire study must be able to be conducted on another occasion and produce the same results. Thirdly, all operationalizations in the questionnaire which aim to say something about the same variable must point to the same findings (known as *internal reliability* or *internal consistency*).

> In contrast to the methods generating qualitative data (e.g. interview and observation), the questionnaire methodology often aims to create statistically correct knowledge of the phenomenon you are interested in –or at least knowledge that you can assume to be generalizable to a larger part of the population than the selection you have covered through the respondents who have answered the questionnaire.

Phase 3 Produce

Following Phase 2, most of the work of reading up on relevant research and gathering empirical material will have been dealt with. You have a prototype (2:1) in which you have made an initial compilation of your empirics and a quantity of text fragments, perhaps sorted into working documents with different kinds of material that you may be able to make use of during the Produce Phase. This is when you seriously start creating your product.

In all probability, there will be parts of the various prototypes from the previous phase which have not yet been completed and which you are continuing with. Not all your empirics are finished yet; maybe you have a number of interviews that still need to be done and feedback from the latest seminar, as well as supervisors who must be brought

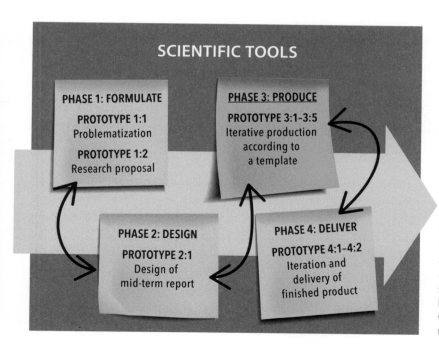

FIGURE 12 **Phase 3: Produce**. During this phase, you will need to choose the template you will be writing your thesis in accordance with. Even though the prototypes you produce during this phase build on what you have done so far, it is wise not to copy too much from your previous prototypes as you will then run the risk of getting bogged down in old thinking. Once again, you may need to recapitulate or even deepen your understanding of the scientific tools in order to better be able to explain what you are doing in your thesis and why.

up to speed. The boundaries between the 4-phase model's various phases are not razor-sharp. But now you are gradually switching over to producing. Exactly how your workflow might look during this phase is described later on. In brief, producing entails creating and presenting a number of prototypes of your thesis in order to arrive, through the writing process, at more and greater insights into what your study will show (see Figure 12).

Arrangement and content of the academic paper

There are different templates for arranging your paper. The two most common ones (Templates A and B) are described in more detail in the section *Two templates for the academic paper* below. The difference between them is how empirics and analysis are presented. While Template A (the separated paper) shows how you can arrange the text if you need to keep empirics and analysis apart, Template B (the interwoven paper) shows how you can arrange them so that you present the empirical material while simultaneously presenting your analysis of it.

> Template A shows how you can arrange the text if you need to keep empirics and analysis apart.

> Template B shows how you can arrange the text so that you present the empirical material while simultaneously presenting your analysis of it.

Both templates pre-require you to have compiled your empirical material in some way prior to starting work on Phase 3. In all probability, you will have compiled your empirics in Prototype 2:1 but may not have analysed them particularly thoroughly. It will not be until you have a version like this that you will finally be ready to decide – in consultation with your supervisor – which template matches your thesis the best.

Maybe the two templates do not tally in every detail with the template used at your department. The headings can have slightly different names and different sequencing, but the main structure exists at the vast majority of universities. Templates A and B both consist of three parts and are usually, as mentioned, described as follows:

Part 1: Describe what you are thinking about doing
Part 2: Do what you said you were going to do
Part 3: Describe and discuss what you have done

Part 1 contains the introduction, literature and theory sections, and method section, and should constitute approx. 30% of the total extent of the paper.

The *introduction* should be written in such a way that you arouse the reader's interest and so that he/she understands why the study you have conducted is needed. This means that you need to begin with a *background description* (which most frequently describes how something is – a circumstance that has been observed and documented somewhere; here, you need to refer to existing research, to articles in the mass media or to other sources you can name). The background description should lead on to the *problematization* in which you allow your thesis to narrow down and focus on the problem that your work deals with. The problematization should have a scientific grounding. This means that the problem you are describing is to be related to and positioned in relation to existing research in the field. The problematization will lead to the thesis's *purpose* and *question formulations* which, against the backdrop of what is in the introduction, become a natural and logical consequence of what has to be investigated. The introduction can subsequently be concluded with any delimitations that may sometimes need to be specified.

Thought model for the preface to the introduction section

Background
This is how it looks in industry/at companies today …
(reference to sources A, B, C .)
Problematization/problem formulation
It is problematic because …
Research in the field says that …
(reference to sources D, E, F .)
We do not know, however …
Purpose
Thus, the purpose of this study is to …
Research question(s)
In order to achieve the purpose of the study, we have to ask ourselves the following question(s) …
Delimitations
The study has not focused on …

In the *Literature study,* you account for previous research into the phenomenon you are interested in studying, while in the section *Theoretical frame of reference,* you describe the theory or theories you intend to make use of in order to understand the empirical material you have collected. The focus of the literature and theory sections should be relevant to what your thesis is about. You are *not* expected to account for all the previous research and theories that you could envisage using in order to understand the problem.

In the *Method section,* you account for what you have done in order to study the problem that you previously formulated in the problematization. You account for your choice of research design, your choice of data gathering methodology, and your choice of analysis method, preferably in this order. You will also need to justify all of your choices on the basis of the purpose of your thesis. When advocating your choices, you can take inspiration from the relevant method literature; it will be appropriate for you to refer, in the method section, to method literature when justifying your choices.

Please note that, in the method section, you are *not* supposed to account for everything you have learnt about method, or all the alternative possibilities that could have been used to achieve the purpose of the thesis. You only discuss possible alternative choices when these, for some reason, are obvious but have been rejected. Do not forget, either, to account for how you have taken issues such as ethics and gender into consideration during the course of your work; what kinds of sources you have used and how you have assessed them and related to them source-critically. This is a matter – like the entire method section – of reasoning as regards what is relevant to your thesis specifically, so that the scientific requirements are met.

Regardless of whether you choose to arrange the final version of your paper in accordance with Template A or B, the middle part should constitute about 50% of the entire extent of the paper.

The concluding part of the paper covers about 30% of the text. It is here that you summarize and discuss your results critically in relation to sources and method. Feel free to begin by briefly repeating the purpose, how you approached achieving the purpose, and what you have arrived at. You can then continue by discussing the study's delimitations and weaknesses, e.g. as regards reliability and validity, and how these can impact upon your findings. After that, you highlight the merits of your work. Bear in mind that you have to argue here why the thesis, despite any shortcomings, is knowledge-enhancing.

Remember to also discuss your contribution vis-à-vis your client. For example, you can provide advice that is based on your findings and discuss the opportunities that these entail for companies (*managerial implications*). You can also make suggestions regarding future research –for both academy and industry.

Prototype 3:1–3:5 (× n)

Below is a proposed workflow in five stages whereby you produce your final product via several prototypes: (3:1–3:5) × n. This proposal will be suitable regardless of whether you are working with Template A or B. This process is very similar to prototyping such as it was described during the previous phases. The difference, however, is that you no longer have the time to be as open as you were to new directions and to new lines of inquiry.

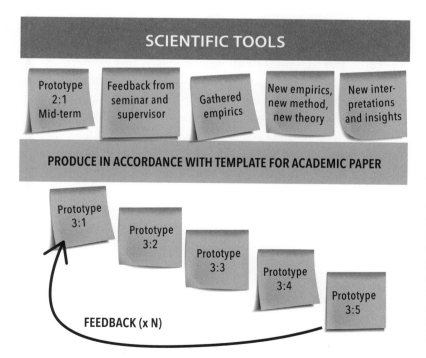

FIGURE 13 Prototype 3:1–3:5, with proposed workflow. Input for phase 3 is everything you have done hitherto, all feedback and insight. Now there is not so much time to conduct empirical investigations or gather new data. We propose a work process of five prototype stages which is then repeated several times (× n).

> **Phase 3 may seem rather messy as what you write in one place in the paper will most frequently entail consequences for what is written somewhere else.**

Phase 3 may seem rather messy since what you write in one place in the paper will most frequently entail consequences for what is written somewhere else. As you analyse your empirics, you may realise that you need to reformulate the problematization, purpose, and research questions. You need to remind yourself that the paper, once you have arrived at your conclusions, has to be argumentative. It may also be wise to think about the *relationship* between theory and empirics. What relationship do you want to put down in writing, now that you have started producing your thesis in earnest? Have you studied your empirics on the basis of a definite idea about what you thought you would find (deduction)? Or have you been open to what the empirical material might say (induction)? What is actually the phenomenon (*explanandum*) that you want to say something about by means of the examples in your study (*explanans*)?

If you are unused to sitting down for long periods writing and producing your own text, then you might find the Produce Phase extra frustrating. This is something to bear in mind when planning your degree project prior to getting started. Our advice is to earmark a proper amount of time for this phase; not least due to the possibility of needing time to allow the different versions of the text to rest during this phase so that you can put the text to one side for a time in order to be able to return to it with fresh eyes.

Often, there are no scheduled seminars during the Produce Phase. In this case, we would like to repeat our advice about creating your own reader groups/peer review groups to which you can present different prototypes of your paper. The help you stand to gain from such a group is almost invaluable, not least because, as we have said several times now, you can hone your thoughts when forced to clarify them and present them to others.

Objectives of work during the Produce Phase:

- recipient-adaptation of the text
- creating alignment (logical consistency) between the various parts of your paper
- deleting unnecessary text.

The workflow we suggest below is something you will have to do in several rounds. Once you have completed one stage, it may be the case that you realise that what you wrote before is not really coherent, and then you will have to do a re-write. The academic paper requires logical consistency. That makes this part of your thesis work difficult. You will have to keep several balls in the air at the same time; at times, you will probably have to reformulate your points of departure completely. This is frustrating and can impact upon your sleep negatively. But, as in all creative work, this frustration leads to the thesis becoming better each time you write a new version – even though it does not always seem so when you are in the midst of all this work.

Before you get going on Prototype 3:1, you should, if you have not done so already, choose which template you are going to use to arrange your paper. Following that – regardless of whether you have chosen Template A or B (or some other template) – you will have to think through which parts of your thesis belong to which part of the tripartite structure for academic papers that is listed below. This tripartite structure will help you to keep the logical consistency in order. In the rest of the description of the workflow, we use this tripartite structure:

Part 1: Describe what you are thinking of doing
Part 2: Do what you said you would do
Part 3: Describe and discuss what you have done

PROTOTYPE 3:1: WRITE AN EXPANDED SYNOPSIS

A good way of starting the Produce Phase is to write an expanded synopsis. This entails producing a document in which you formulate, under all the headings that your template prescribes, a brief description of what will come under each respective heading. If you have gathered qualitative data, you can try to summarize your spontaneous impressions while gathering a number of brief formulations under the heading in Part Two, where the findings and analysis will later be written. If you have gathered quantitative data, this will be difficult; in this case, you should instead write what you envisage as coming under the different headings.

Put simply, writing the synopsis means that you

- write explanatory texts under each heading, known as *ingresses*. Describe in each ingress what the passage is about. Ensure that

you find the continuity between the ingresses so that all parts are cohesive.
- think through the various parts of the paper in relation to how you have briefly described each section.

Beginning each new prototype attempt during Phase 3 by writing an expanded synopsis is similar to the challenge of switching between divergent and convergent thinking (see Figure 6, p. 32). A new synopsis entails taking a step back (diverging) and thinking about the whole. The synopsis is something you will be able to use as an initial draft of your abstract; i.e. for the summary that you will write at the very end of the Produce Phase. The synopsis will also be valuable to you during the rest of the Produce Phase as it will function as a map; i.e. a document that you will be able to return to in order to find your way when you are at risk of getting entangled in the tricky details of the various texts you are writing under each heading.

We thus suggest that you use your expanded synopsis as a foundation document right the way through Phase 3. As you will see, Prototype 3:2–3:5 entails developing the synopsis using texts under each heading in order to thus gradually build up your thesis.

PROTOTYPE 3:2: WRITE A VERSION OF THE SECOND PART OF THE THESIS

Once you have written an expanded synopsis, the best thing will be to start in the middle part and work through Part Two. It is now that you will be writing down your findings. How this is achieved will depend on which method of analysis you use. Once you have formulated a text describing what your empirical study shows, you will have to think about what these findings entail in relation to, and against the backdrop of, previous research. What does you study say that no other study has said? What can you show that has not been shown before? How does your study shed light on the phenomenon that your thesis deals with?

This prototype is about gradually working up the answer to your research questions. When you have found the answers, you must formulate, with the support of your empirical study and previous research, arguments as to why your answers are tenable (valid). Use the headings which you decided upon and which will strengthen the line of argument for your results.

- Formulate the text that will come under each heading. Move, delete, and write new text.
- Work hard at substantiating your arguments using theory, empirics, and previous research.
- Also ensure that you remove all the unnecessary bits which do not help the reader to understand and which do not bring added value to your client. You have to think about which arguments have to be substantiated extra carefully and which do not require an equally strong underpinning.
- Return to your source-critical discussion and to your critical discussion about research method. When presenting your argumentation and analysis, you will have to honestly tell, and remind, the reader if some of your findings have weaknesses due to the nature of the sources or the design of the methodology.
- Adapt the text to the reader and create logical consistency. A good way of creating consistency is to write *ingresses* under each heading to inform the reader of what you will be dealing with in what is to come. Also add *part summaries* of what you have written prior to starting a new section. These ingresses and part summaries can be deleted from the last version if they are not needed. But during the Produce Phase, they are important because you yourself will need to keep tabs on what you are doing. The part summaries will also help you to write your conclusion.

PROTOTYPE 3:3: WRITE A VERSION OF THE THIRD PART OF THE THESIS

Once you have completed a version of Part Two, it will be a good thing to work with the end of the paper. Write out the findings you have arrived at, i.e. the answers to your research questions on the basis of the knowledge you have acquired by means of reading literature and doing empirical studies.

Investigate whether or not you have evidence for your most important findings. It will be appropriate for you to use your findings as a governing logic when working on the second part in the next prototype.

- Have you answered the questions that have been formulated?
- Is an important piece of the puzzle missing? If the answer is yes, you will have to choose, in consultation with your supervisor,

whether to do further empirical investigations or to change your purpose and research questions so that they match what you have arrived at. In most cases, there will be no time for further research and you will have to adapt to what you have had time to do.

> **Many students think that they are not allowed to change their problematization, purpose, and question formulations. But that is wrong.**

Many students think that they are not allowed to change their problematization, purpose, and question formulations. But that is wrong. In all research, points of departure are changed and introductions are reformulated. The findings arrived at almost never tally with the points of departure used at the beginning of the research project. New insights from theory, method, and empirics constantly surprise us. If we are not surprised, we are either dogmatic or we have a trivial point of departure (as if it is not a good idea to investigate a certain issue as we already know what we will find).

Writing new text is wise as there is a great risk that you will lose your focus and produce poorer results if you cut and paste from old texts.

It is in this situation that you will find shortcomings and inconsistencies. Maybe a number of interviews are missing, for instance, or the theory you have in place in order to explain your empirical observations is insufficient.

PROTOTYPE 3:4: WRITE A NEW VERSION OF THE FIRST PART OF THE THESIS

Now that you have arranged things so that Parts Two and Three are logically coherent, it is matter of rewriting Part One.

- Is your *literature review* correct and does it match what you have done?
  Are you specifying the right theories in your review, i.e. the ones you will be using later on? Have you used all? Should any be removed? Have new concepts been added that you need to explain? It is not uncommon to realise here that you need to do some further reading, and that some of what you read during previous phases is irrelevant in relation to what the thesis has now come to deal with. It can feel difficult to delete sections you have worked hard at in previous versions of the literature and theory sections;

however, if they add nothing to Part Two of the paper, they will have to go.
- Proceed by thinking about the *method section*. Does what is written there tally with what you have actually done? Clearly justify your choices using references to suitable method literature. Also return to your source-critical discussion and think about how and to what extent you need to discuss source-criticism in your method section.
- Next review your *problematization*. Have you really been working on the problematization you have specified, or does it need to be reformulated? Does it take up and clarify all the concepts that the purpose and research questions contain? Also check the purpose and research question against what you have actually done. Does the purpose tally? Is it concisely written? Do the research questions match the purpose and what you have actually done? Are you answering all the questions? If not, remove the ones not being answered and add any new ones emerging during the course of your work.
- Do your *delimitations* tally or has your research field changed (shrunk or expanded)?
- *Delete everything that is unnecessary!* It is very common for the first part of the paper to be overly long – the thesis becomes "front heavy". Regardless of which template you use, the reader quickly wants to get to the really exciting bits: i.e. how you have dealt with and analysed your empirics using the theory, and which conclusions you have drawn. It will not be until then that fresh knowledge and new insights arise. You must not write long accounts of all the imaginable methods and theories existing in your scientific field. Your supervisor knows this already, and your client is not interested. The paper is not in existence for you to account for and report upon the fact that you have read loads of research literature. You have to fulfil the purpose of the investigation and answer your questions in the most effective way possible.

PROTOTYPE 3:5: WORK THROUGH THE LANGUAGE, LAYOUT, AND FORMALITIES

You have now made an initial attempt, and you have an initial version of the text. The final stage –before starting up again (!) – is working through the language, layout, and formalities (title page, abstract, list of contents, and references).

- Use the spelling checker in your word processor. Make sure that you are not using slang or clichés, and that all words and terms are being used in a conscious and consistent way.
- Make sure that you have written complete sentences and that your division into paragraphs is correct. Review the transitions between various passages and between the various sections of the paper. Do they naturally lead on from each other language-wise, too? Can the headings be changed in order to specify, in an interesting way, what is to come? Just because you have used a certain template, it is not self-evident that you have to use that template's headings. It is not unusual, for instance, to have your own, content-related headings. More tips and advice on good scientific style may be found on p. 50.
- Review the layout of the paper, too. At many seats of learning, there is a digital template that you must use; make sure that you comply with this. Think about whether or not you can visualize your text in a way that helps the reader to understand what you are writing. Make sure that tables, diagrams, figures, and other illustrations are numbered, have headings, and are explained in the body text.
- You will also need to review the references in the text and make sure that all references are included in the reference list and that this has been set up in the correct manner vis-à-vis the reference management system you have chosen.
- Once you have finished with the text, after the last prototype, you will also have to create a title page on which the heading of the thesis, your name and the name of the seat of learning are all listed, as well as an abstract that functions as a summary. Also make sure that you update your list of contents.

> The final stage is working through the language, layout, and formalities.

MORE ITERATIONS – MORE PROTOTYPE ATTEMPTS

We suggest that you work through the prototype stages of Phase 3 at least twice, maybe more. But before starting on a new iteration, a good tip is to let others have a read; during this time, you should take a break from your work for a few days. Let the thesis rest. It is easy to become blind to your own writing when working intensively on it. Thus, put it to one side for a time and return to it later on to read it with fresh eyes. This requires a certain amount of planning so that you have time before the next phase starts. However, we can promise you that this pause will be profitable. Yet another tip is to intersperse your writing with oral presentations.

Rewriting and working through the paper in the manner described above may seem unreasonable in comparison with your previous experiences of producing text at university. However, we claim that it is entirely impossible to achieve the completion of a degree project with the very first version. You will have to continue prototyping, iterating, deleting, moving, and adding. During the course of this work, you will change the problematization, purpose, and question formulations. You will work on creating logical consistency between the various parts of the thesis and, in doing so, gradually deepen your analysis. This is how scientific work proceeds. Even though it is not scientific work in the same way that a degree project is, we allow Figure 14 to illustrate our argument. This is a screen dump of the process we underwent while writing this book. Over a period lasting

> We claim that it is entirely impossible to achieve the completion of a degree project with the very first version.

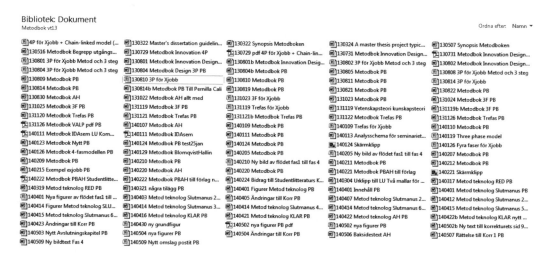

FIGURE 14 List of some of the different prototypes of this book during its genesis.

more than a year, we wrote a number of different versions and, during the course of this work, made several radical changes.

The iterations will probably not just make the text better, they will also provide you with renewed insights into what the thesis is all about. Remember that the final prototype during Phase 3 will not describe or reflect your work process; it will advocate the findings you have arrived at.

> **Insights during the final prototype attempts**
>
> A couple of students had been commissioned by a major IT consultancy firm. The investigation was about rationalizing collaboration between different departments with the aim of being able to offer added value to the customer while reducing costs. The students arrived at the fact, among other things, that the corporate culture of the IT consultancy business was an important factor in this complex of problems. During the final stage of the writing process, during Phase 3, they discovered that their final part was not at all coherent with the introduction and argumentation. It turned out that they had not defined the concept of culture sufficiently clearly. The students, and their supervisor, had become situationally unaware and did not see that the concept of culture was both wobbly and imbued with different meanings in different parts of the thesis. The matter was tolerably rectified, but time was of the essence and a number of unclear points remained, something which was also pointed out by the principal examiner at the final seminar. These students were not awarded the highest of grades.
>
> The lesson that can be learnt from this is that it is not until you bring together the various parts of the thesis that you can see the whole and the continuity, or lack thereof. Make sure, in good time, that you get an outsider to read, with fresh eyes, one of the very last prototypes. Then you will have time to correct the unclear points that both yourself and those most immediately concerned are no longer able to see.

SCIENTIFIC TOOLS

Two templates the academic paper

Templates A and B are really not as different as they may seem at first glance. The biggest difference is the fact that, in Template A, you split Part 2 of the paper into an empirical account and an analysis account. In Template B, you combine the empirical account with the analysis account and sort the text under headings adapted to the arguments for your findings.

Template A: The separated paper

This template is common when a degree project is based on quantitative material or when, for various reasons, you want to differentiate the presentation of the empirical findings from the analysis of these using the chosen theory. You may, for instance, want to account for the empirics in chronological order and discuss them thematically. It may also be wise to start writing like this – even though, in a later version of the text, you choose to follow Template B.

Template A: The separated paper

- Title page (title of thesis, author's/authors' name(s), course/group, seat of learning/department, year)
- Summary/abstract (should be approx. half a sheet of A4 and contain a brief background description, purpose, research question, findings, and recommendations to client; most often written in both English and a local language)
- List of contents

1. Introduction
2. Literature study
3. Theoretical frame of reference
4. Method

} Corresponds to **Introduction** in Template B

5. Empirics
6. Analysis and conclusion

} Corresponds to **Argumentation and analysis** in Template B

7. Discussion and recommendations
8. References

} Corresponds to **Conclusion** in Template B

The *empirics section* is where you account for your empirics; however, most of the time, you will not be able to present your empirical discoveries just like that. Neither will you be able to sort the empirics in accordance with where or how they have been gathered (by interview, company or similar). The empirics, even though they are accounted for in a separate part, must be sorted in a well thought-out manner – thematically, chronologically, or similar. Then follows the *analysis and conclusion section* where you discuss your empirical observations using the theoretical framework you accounted for in the theory section. Finally, you conduct a concluding line of reasoning in the *discussion and recommendation section*.

Template B: The intertwined paper

Template B does not have a separate part in which the empirics are accounted for separately followed by an analysis assisted by theory, as in Template A. Instead, in Part 2, you will argue on the basis of both your empirical discoveries and your theoretical terms, as well as the findings of previous research. This places great demands on how you arrange the argumentation and analysis section.

Template B: The intertwined paper

- Title page (title of thesis, author's/authors' name(s), course/group, seat of learning/department, year)
- Summary/abstract (should be approx. half a sheet of A4 and contain a brief background description, purpose, question formulation, findings, and recommendations to client; most often written in both English and a local language)
- List of contents
- Part 1. Introduction ⎫
- Part 2. Argumentation ⎬ How the subheadings and analysis are formulated will be determined by the specific thesis
- Part 3. Conclusion ⎭
- References

The completed argumentation part should not have the empirics sorted in accordance with where or how they were gathered – that is, not by interview, company or similar (even though this is an initial suitable step following completion of the empirical study; exactly at the start of Phase 2). Instead, sort by *categorizing* or *thematizing* your empirics.

Use the categories or your themes as headings in the argumentation section. An effective way of arranging the argumentation part is to use different parts of your findings as headings/sort logic, advocating your findings, under each heading, using empirical discoveries, theory, and previous research.

In Template B, you will thus present your analysis while simultaneously presenting your empirics. Use your theoretical terms to explain how you arrive at your findings. This entails answering the questions you posed by way of introduction in the argumentation and analysis part, while simultaneously discussing them in relation to the theoretical terms of relevance, and in relation to findings from previous research.

The arrangement of this part and how the subheadings are formulated greatly depend on the investigation's purpose, research question, and method, but also on what kind of empirics you have. The arrangement must always be such that it supports your argumentation and highlights your findings.

> **Findings as the logic behind the arrangement**
>
> A degree project author investigated how a company would approach selling services aimed at environmental sustainability. Her solution was to recommend a process of four stages: Screen, Evaluate, Build, and Plan. These four terms became her chapter division and sort logic in part two of her paper (she used a modified Template A). Under each heading, and in support of each and every one of the four stages of her process model, she set forth her empirical discoveries, her theory, and findings from previous research.

Template A or B?

Choosing between the two templates A and B is dependent on which template your university, department, and supervisor prefers, as well as on what suits your thesis. In the main, all templates contain the same components.

Template A is clear, with its distinctive chapter division via which the headings precisely specify what all the parts contain. Even if, at a later stage, you choose Template B, it might be a good idea to start writing in accordance with Template A as you will then be making sure to keep empirics, theory, and your own interpretations apart. It may be difficult, especially for your readers, to know what is what, in any event if you are not very clear regarding where what you are writing in the analysis and conclusion section comes from. At the same time, there is a risk that Template A may lead to the separation and modularization of the thesis. Our experience is that a paper which is arranged in accordance with Template A is not always coherent as students are tempted to paste in texts from previous prototypes (e.g. 1:2 Research proposal and 2:1 Mid-term report). This involves the various sections on literature, theory, method, and empirics becoming descriptive instead of argumentative and analytical, as well as the various parts not being coherent. There is also a risk of the empirics being presented as *raw data*, something which can be appropriate in certain cases, but not always. Often, there is no reason, in the social science fields, to sharply differentiate between an empirical account and an analysis.

Template B's strength is that it encourages *alignment* – logical consistency – between the various parts of the thesis, prototyping and the preservation of inquisitive uncertainty, the deletion of unnecessary information, recipient-adaption, and clear and well-founded argumentation of the paper. Template B, however, also has its disadvantages. It may be difficult to reduce the introduction when you have really worked hard at reading up on theory and method. The argumentation and analysis part can be perceived to be diffuse and difficult to divide into different chapters. It can also appear to be in breach of lessons learnt from previous theses or smaller reports, where you have been told, perhaps, that the empirics are to be presented in an objective and raw manner. Additionally, you have to be very clear when following Template B regarding who says what: which statements originate from the empirical material, which originate from the literature, and what your own analysis is. If you do not succeed in keeping this apart, the argumentation and analysis will be messy and difficult to read.

ON THE ANALYSIS OF EMPIRICS

The empirical material you have gathered cannot be presented purely in its raw state. It has to be processed and this processing has to be carried out in a critical manner – you should have an open but reflective approach to your material.

During this work, it may be wise to keep *reflective writing* and *presentative writing* apart (see pp. 51–52). Most of the text you produce as a result of reflective writing will not be included in the final thesis. Too one-sided a focus on presentative writing often inhibits reflective writing, and thus the development of ideas.

> During this work, it may be wise to keep reflective writing and presentative writing apart.

If you are working with *qualitative data,* you will often be working according to an inductive or an abductive process whereby, in parallel with your analysis work, you read up even more on literature, with your understanding of the material frequently increasing though the interaction between existing theory and your own empirics.

If you are working with *quantitative data,* you will be working in accordance with either an inductive or a deductive process and, depending on which your analysis work can differ somewhat in appearance. If you are working inductively, you will need to be prepared to a greater degree to attempt to analyse the empirical material in different ways, by conducting univariate analyses, where you look at one variable at a time, bivariate analyses, where you place two variables in relation to each other, and – when doing more advanced quantitative studies – multivariate analyses, in which you investigate how several variables relate to each other. If you are working deductively, the analysis methods will to a greater degree be given by the hypothesis/hypotheses you have initially formulated.

In what is to come, we run through four different ways of analysing the gathered material:

- thematic analysis
- narrative analysis
- discourse analysis
- quantitative analysis.

Following that, there is a section on sources, i.e. on the origins of the empirical material, and on how you, as a degree project author, can (and should!) relate to these in a scientific way. There, you can also read about reference management. Finally, we describe how the text you are prototyping when you are finished with your analysis has to be proposition driven (which is not the same as the purpose), and how the text must therefore be argumentative.

Thematic analysis

A very common way of analysing qualitative empirics in the social sciences is to make a thematic analysis. This entails the use of categories into which you sort your empirical material in order to then, on the basis of the categories, answer your question formulations.

Exactly which categories you use can vary. Sometimes, it can be a matter of using categories that you find in the material, often in the form of the words and terms that the people you have been studying make use of. It can also be a matter of themes that emerge when you read through the material – themes that you come across yourself. Or, it can be a matter of categories that you find in the existing literature in the field. The very least refined form of category can be similarities/dissimilarities in the empirical material; for instance, you can arrange the material by company or by informant.

Once you have decided which categories you want to use to sort your material, you will need to carry out the sorting itself. This entails connecting the different parts of the material to the categories.

> Thematical analysis entails the use of categories into which you sort your empirical material.

"Materiality" in the digital world

What is *materiality* in the digital world? In order to study that question, a student conducted an observational study of a software development company. Out of the empirical material that he collected, five themes crystallized:

- the use of physical artefacts
- the reification of digital materiality
- connections and associations
- the selection of digital materiality
- digital materiality in digital innovation.

Under each theme – which also worked as headings in his paper (it was arranged according to Template B) – he conducted a developed line of reasoning. After that came a final discussion in which, on the basis of his themes, he advocated the fact that digital materiality has a double function by which means it creates associations between, as well as connects, people and artefacts, but that it also, specifically by means of this function, "comes into being".

A concrete way of doing this is to give the different categories different colours and then colour the different parts of your material on the basis of which category it belongs to.

Once you have connected sentences, paragraphs, and other parts of the empirics to the categories, it is time to look at each category individually. It will be appropriate to read through all the material belonging to a category and to think about – preferably during reflective writing – how to understand each category. What can be said about the respective category? How does it answer your question formulation? This process of reduction sometimes entails having to reformulate your research question which is more usual than unusual when working inductively.

By reflecting upon what the different categories entail, you reduce your material and become increasingly ready to write a text under each category in which you present what that category says. In this presentative writing, you should imagine that you are advocating what you wish to call attention to, under the specific category, in relation to your overarching problematization.

Narrative analysis

Another way of analysing qualitative data is by doing a narrative analysis. Narrative analysis is based on the basic assumption that storytelling and stories are one way for us humans to understand the world. You can conduct a narrative analysis in two ways: either by organizing your material in the form of a story or by finding and analysing any stories in the material.

Organizing the empirical material in the form of a story entails arranging the material so that there is a beginning, a middle, and an end. The simplest form of narrative analysis is arranging the material in chronological order. Then, the important thing is to do this in a conscious way: thinking about why you are allowing the story to begin and end where it does, who you have as the main character(s) of the story, from whose perspective you are telling the story, and which episodes you make the story consist of. The story is to be told in such a way that it shows something – don't just compile the material chronologically. You can also arrange the material in accordance with some other form of logic, e.g. cause–effect or input–output. These types of narrative analysis are suitable when you have a lot of data from a number of different sources and types of data gathering methods

> You can conduct a narrative analysis in two ways: either by organizing your material in the form of a story or by finding and analysing any stories in the material.

(when, for example, you have conducted an ethnographic study and have observational material, interviews, and documents).

When sorting the material in the form of a story, you reflect upon what the story says in relation to your research questions and existing literature (preferably in the form of reflective writing), and then you link what you see to this.

Finding stories entails going through the empirical material looking for any stories in it, e.g. in interviews or documents. You then analyse these, for instance by looking for the hero, the anti-hero, the peripeteia (the turning point of the story), and so on. You can also compare various beginnings, i.e. where different stories start, as a way of understanding how different informants construct the reason for the phenomenon being talked about. Here, too, reflective writing can work as a good way of progressing in your thoughts as regards to what this says in relation to the question formulations and the extant literature in the field.

Discourse analysis

Discourse analysis is based on the assumption that language not only expresses but also creates and changes the phenomenon that we are speaking of while using language. A not uncommon form of discourse analysis is *critical discourse analysis.* In addition to the basic assumption that language creates conceptions, critical discourse analysis is based on the assumption that language, in doing so, reproduces or questions underlying ideologies. This means that a critical discourse analysis takes an interest in what the discourse is trying to achieve, and in which means are used for that purpose.

When conducting a discourse analysis, you put questions about this to the text. What is the discourse trying to achieve? How is the discourse achieving this? What are the fundamental statements in the text? How is the empirical material (the interview, the document and so on) persuading me as a listener/reader? How are the ideas being justified? How does the empirical material assure me that what is being said is true? In which ways are the ideological basic assumptions being reproduced or questioned?

> **"Sustainability" in annual reports**
>
> A student was interested in studying how the notion of sustainability appeared on the agendas of quoted companies, and the connotations of "sustainability" at these companies. He downloaded from the Internet all the annual reports of 10 quoted companies, from 10 years ago up until today. Following that, he focused on the letters written by MDs that are often to be found in these annual reports and analysed how the word sustainability was used in these different texts. His analysis showed that sustainability was largely associated with environmental issues and finances, which are just two of the four dimensions raised by the Brundtland Commission's definition of the term. The social and cultural sustainability dimensions were never touched upon in the MDs' letters in these annual reports.

In discourse analysis, reflective writing is a good tool when it comes to answering questions; here too, the result is presentative writing when you have worked out how to answer your question formulations in relation to what the material says.

Quantitative analysis

Quantitative analysis is suitable when you have collected quantitative data. If you have used a questionnaire, the first thing to do is to code the data you have collected. This entails giving the questions and answers different numerical values which are subsequently used to describe and analyse the empirics statistically (see Figure 15). This is frequently done with the help of some form of digital tool, e.g. Excel or SPSS.

Before you start analysing data, you will need to ensure that all the respondents who have answered belong to the population (see p. 63). This is done by going through the questions you have posed with regard to age, gender, or anything else characteristic of the population. If a questionnaire contains extreme values in some way; that is, if the answers of a respondent deviate very much from other answers, you take it away. Such a questionnaire is called an *outlier* and is not deemed representative of the population (since it is extreme), instead just providing incorrect input into the analysis.

There are different forms of analysing quantitative data. What you choose to do depends on the problematization, what purpose you have

> How you choose to analyse quantitative data depends on the problematization, what purpose you have, and how advanced an analysis you have the possibility of doing.

with the study, and how advanced an analysis you have the possibility of doing.

The simplest form of quantitative analysis entails doing a *univariate analysis* in which you study one variable in the investigation at a time and make a statistical description of this, for instance by looking at variation and average. How this is done will depend on what kind of variable you use.

If you start out from a *nominal variable*, a variable in which there are more than two alternatives that cannot be ranked, you can demonstrate variation by showing how many have chosen which of the categories the questionnaire has provided, and averages by means of calculating the most frequent value among the answers submitted (*frequency*).

On the basis of an *ordinal variable*, i.e. a variable in which the answers can be ranked but the answers have unequally large distances between them, you can show the *median*, the answer in the middle of all answers (where 50 percent of the answers are above and 50 percent below), and the variation in the material by illustrating the spread of the answers. One way of clarifying the spread around the median is to draw a boxplot.

If you use a *ratio/interval variable*, i.e. a variable in which there are more than two alternative answers that can be ranked with an equal difference between the answers, then you will be able, on the other hand, to calculate the mean by working out the arithmetic mean value and the *variance*, the standard deviation from the mean.

It is important to remember when doing a univariate analysis that, when using one of these, you will not be able to explain any variance; you will only be able to point out that a variance exists. You will not get very far in your analysis using univariate analyses.

A more advanced form of quantitative analysis is *bivariate analysis*, in which you juxtapose two variables against each other. Frequently, one of the variables is called X while the other is called Y, and it is not unusual to draw these in a diagram where X is on one of the axes and Y on the other, e.g. by drawing a boxplot, a scatter plot, or a table in which you show how many respondents there are in each respective category in each variable.

The most advanced form of quantitative analysis is *multivariate analysis*, where more than two variables are set in relation to each other. There are several different methods of doing multivariate analysis, e.g.

1. How many employees work at your company?
   __200__ employees   NUMBER OF EMPLOYEES = RATIO/INTERVAL VARIABLE

2. Which of the following alternative best describes your operations? (mark one alternative)
   [ ] Wholesale and retail
   [ ] Real estate company and property management
   [x] Information and communication
   [ ] Transportation and storage
   [ ] Legal, financial, architectural, or technical consultancy and R&D
   [ ] Financial and insurance
   [ ] Education, care, and social services
   [ ] Leasing, job agency, travel services, other support services
   [ ] Personal and cultural services
   [ ] Hotel and restaurant
   [ ] Advertising, market research, other operations in law, finances, science _____
   [ ] *Other (please specify)* _____   TYPE OF SERVICE COMPANY = NOMINAL VARIABLE

3. How often do you introduce a new service?
   [ ] Monthly
   [ ] Quarterly
   [ ] Biannually
   [x] Annually
                  FREQUENCY OF INTRODUCTION OF NEW SERVICES = ORDINAL VARIABLE

4. Approximately how many man-hours per week do you dedicate to developing new services?
   __5__ hours/week
            MAN-HOURS DEDICATED TO SERVICE DEVELOPMENT = RATIO/INTERVAL VARIABLE

Data compilation:

Variable number →

01	02	03	04
200	3	4	5
33	2	3	1

Answers in questionnaire → (row 1)
Answers in questionnaire → (row 2)

FIGURE 15 Coding the answers to a questionnaire in Figure 11. The example shows how the answers in questionnaires are firstly coded (i.e. the answers are translated into a number) and then compiled in a table. This is done for all the questionnaires; here, a second questionnaire has generated data that exists in the compilation. Frequently, this is not done manually, but the values are fed in to some kind of software, e.g. Excel or SPSS, in order to simplify the analysis. Some digital questionnaire tools also generate data compilations.

regression analysis, by means of which you try to understand how one specific variable (known as the *dependent variable*) changes when another (*independent*) variable changes (while a further independent variable remains unchanged), or *factor analysis*, which is a way of finding connections between different variables by means of looking at variations in these variables purely mathematically.

For assistance in processing and analysing quantitative data, you often use some form of digital tool such as Excel or SPSS. Note that it takes a lot of time to learn these tools to the extent that you can utilize their functions for what you want, something that is wise to bear in mind when making a plan regarding your degree project. On the other hand, as an engineering student, you have most probably read a sufficient amount of maths and statistics in order to relatively easily manage to learn how to do multivariate analyses, and if a quantitative study is suitable for your thesis, you should absolutely consider conducting such a study.

ON SOURCES AND REFERENCE MANAGEMENT

You have to evaluate the reliability of your empirical sources and investigate whether or not the facts and statements in the source material are credible. In other words, you have to make use of so-called *source criticism*. It is also important, from the point of view of credibility, that you clearly specify – refer to – the sources you are using. You have to give your client or opponent the possibility of easily being able to find and scrutinize your sources.

Primary and secondary sources

There are many ways to classify and categorize sources within different scientific fields. Within some scientific fields, everything included in an investigation is called a source, both empirics and literature. Other research specializations instead talk about primary data (things you have found yourself) and secondary data (things gathered from other studies).

In this book, we use what is usually known as *a functional source concept*. In accordance with this, there are no sources/materials which *in themselves* are either primary or secondary. There is no inherent characteristic of the source that determines its categorization. Division of the source material must instead be done relative to the specific study, and the division between what is a primary source and what is a secondary source depends on their functions within the study. You will have to decide yourself, on the basis of your purpose and research question what are to be regarded as primary and secondary sources.

Primary sources are the empirical material that you yourself produce and/ or material that is very close to the research object and the purpose of the investigation. This applies regardless of whether the source material is oral, written, or in some other form. Examples of primary sources include interviews with actors who have intimate knowledge of what you are studying, written material published by a company or organization, your own interviews with representatives and experts in the industry or field you are studying (in cases where these have specific knowledge of your research object), and quantitative material closely linked with your object of investigation (whether or not you produced this material yourself).

Secondary sources are information that does not have an equally clear proximity to your object of investigation. For instance, this can be previous research into the company or industry, interviews (which you have either conducted yourself or taken from periodicals with researchers in or representatives of the industry, who make general pronouncements on the field), and quantitative material that describes the object of investigation's wider context (whether you have produced this material yourself or not).

Putting more time into thinking about how to categorize your sources is not some old academic rule. Categorization has to do with the primary and secondary material you are using in your research needing to be able to be scrutinized and evaluated. Since you are basing your findings on the sources, it is important that they are used in the right way. And, in accordance with the functional source concept, no source in itself is bad or worthless. Value depends on which questions you pose and whether or not the source can provide answers to your questions. As a researcher, you must advocate this yourself.

When evaluating different sources, the question of reliability and validity crops up again. You have to be certain, you see, that the information and data you produce via your investigation provide reasonable answers to the questions you have asked. One source can have a high level of reliability, e.g. information about Swedish exports and imports from Statistics Sweden.

> You will have to decide yourself what are to be regarded as primary and secondary sources.

> In accordance with the functional source concept, no source in itself is bad or worthless.

> **Primary and secondary sources**
>
> Student A was working with a degree project dealing with how companies work with knowledge transfer during retirement transitions. Student A chose to do a comparative case study of four major knowledge-intensive companies and collected data by means of conducting interviews. In order to obtain background facts about the companies, this student also read their annual reports. Student A made a thematic analysis of the empirical material. The interviews are, in this case, to be regarded as primary data. The annual reports are to be regarded as secondary sources (since they do not describe the knowledge transfer processes at the companies directly, but provide background information about the companies).
>
> Student B's degree project dealt with how major listed companies build their image as being socially responsible. The empirical material consisted of the annual reports from the ten largest listed companies in Sweden over the past five years. This student did a discourse analysis of the annual reports. The annual reports are thus to be regarded as primary sources (since it is these that the student is analysing).

Validity, however, is not simply linked to the information being correct and to our trust in the source. The validity of the source depends on whether or not it matches the investigation's purpose and your research questions Export/import statistics can be used in many investigations, but not in all, of course. One might say that reliability is a necessary but insufficient pre-requirement in order for a source to be useful.

The fundamental questions that you must put to your empirical material are whether the sources are reliable (whether they are correct, produced using the right method etc.) *and* whether they are valid (whether the material can provide answers to your questions).

Source evaluation and source criticism

In order to help you evaluate the reliability of your empirical sources – regardless of whether they are primary or secondary– you make use, as previously mentioned, of source-criticism. You must critically analyse and discuss whether or not your sources are scientifically useful to your purpose. We have processed the most common source-critical criteria a tad in order to fit a general social science environment better (see the checklist on the next page).

Checklist for source evaluation and criticism

Authenticity
- Who is the author of the source? Is the informant an expert in the field and neutral? Can the information be checked? (Does the author, in turn, clarify his/her sources, or does he/she account for an approach which he/she has used to arrive at his/her conclusions in such a way that the investigation or experiment can be repeated?)
- Have others approved the publication and checked it factually?

Proximity and dependence
- Is the information current? (Is there, for instance, a date of publication?)
- Does the informant have any intimate and deep knowledge/insight regarding what is being stated?
- Is this first-hand information, or is it second- or third-hand (rumours)?
- Is the information from the source in a state of dependence upon others; is the source handing over information without having any intimate knowledge of the subject?

Tendency
- Does the source contain tendentious information which depends on the will of the informant to portray the matter in a certain manner? (Is an authority or major organization behind the information, and is it important to the organization that the information is correct? Is the source a research publication that has been reviewed by other researchers? Is the source a textbook published by an academic educational aids publisher?)
- Should financial, political, or power factors influence how you evaluate a particular source?

Representativity
- Is the material in the source representative of the phenomenon or group of actors you are investigating?
- Are the statements/information in the source significant and commonly occurring or are they unimportant, peripheral, and odd? (Do other sources say the same thing?)

PLEASE NOTE: THE SAME CRITERIA APPLY, BY EXTENSION, TO VERBAL SOURCES.

SOURCE: ADAPTED FROM ROLF TORSTENDAHL (1971), *INTRODUKTION TILL HISTORIEFORSKNINGEN: HISTORIA SOM VETENSKAP*. STOCKHOLM: NATUR OCH KULTUR.

A simple way of analysing and evaluating the empirical material you have gathered is to point out who the *sender* is, what the intended *message* is, and who the envisaged *recipient* of that message is. Thinking about the sender, recipient, and message can help you to see that your informant, during an interview, may have an agenda of his/her own as regards what he/she says. Maybe you, as a researcher, are not the only recipient of the message. The same thing applies to documents that you collect and use as primary or secondary sources. The sender, the company or organization, may have a special recipient in mind and may thus have adapted the message accordingly.

> When doing a source-critical evaluation, it is wise to observe the journalistic principle of several concordant sources being better than one.

When doing a source-critical evaluation, it is wise to observe the journalistic principle of several concordant sources being better than one. The same principle is used in court whereby the prosecutor's case becomes stronger and stronger with each witness, as well as with any other evidence, circumstantial or otherwise, that can be presented. However, the principle of "the more, the merrier" does not always apply. Sources and witnesses can lie (authenticity), have a poor proximity and knowledge, and can be influenced by each other or by public opinion (proximity and dependence). Furthermore, they can also have their own motives for distorting their statements (tendency) or may not be at all representative vis-à-vis the subject being dealt with (representativity). The four criteria in the checklist must be used to determine a source's reliability and validity.

During your degree project, you should thus attempt to test your source material on the basis of the above-mentioned criteria. However, it will not be enough to mention that you have been source-critical during the introduction and then completely forget about it. Later on in the paper, when discussing your findings, which are based on your interpretation of the sources, you will have to return to the source-criticism. It will not be enough to initially state that some of the source material is of low validity and then use the same source as if it were a strong source of high validity. In this latter requirement regarding implemented and consistent source-criticism, seats of learning may differ. This requirement can also be variously and strongly expressed on the Bachelor's and Master's levels. Consult with your supervisor on this issue.

It is often a fundamental rule not to use Wikipedia, or any other

Internet source not having a clear author, as a primary source during a scientific investigation. This is not primarily because of reliability per se. Wikipedia can be at least as correct as, for instance, a printed work of reference. Not being able to use Wikipedia as a primary source is due to uncertainty regarding provability. In scientific work, someone else must be able to check the source's reliability and validity, and this can be difficult as regards certain Internet sources as the information on the website may have changed since you last retrieved it. The latter is especially valid in controversial subjects where information may sometimes change several times over. Wikipedia is, however, an excellent resource of knowledge. Often, references may be found there which you can use to obtain ideas regarding where to continue reading.

Reference management

The fundamental rule regarding references is simple: Make sure that you provide, in a logical and instructive way, as much information as will enable a reader/opponent to easily check what is being claimed in the thesis. If you give your client advice, you must be able to show that you have well-underpinned arguments for this. Specifying references is to be seen as your insurance as a researcher or consultant. If something is incorrect, you will be able to show that you are not at fault. Referencing also serves the purpose of avoiding any suspicions of plagiarism. It is absolutely forbidden to make use of other people's texts or research findings without specifying where you have taken the information from. Passing off someone else's findings as your own is plagiarism. Neither are you allowed to reference something, e.g. a book, which you have not read. If you find information in a book that references another book, then both must be referenced so that it is clear how you obtained the information yourself – but only authors, books, or reports you have actually read need to be included in the reference list at the end.

At the end of the paper, there must be a *reference list*. This is to be arranged in such a way that the references are sorted under different headings, e.g.: Primary/secondary sources; Interviews/verbal sources/written sources; Material from the Internet; Literature and adaptations, and so on. The arrangement is something that you decide on yourself in

> " The fundamental rule regarding references is simple: Make sure that you provide, in a logical and instructive way, as much information as will enable a reader/opponent to easily check what is being claimed in the thesis.

Two different ways of specifying references in running text and in the reference list – the Harvard system and the Oxford system

Harvard system	In running text	In the reference list
	The 4-phase model is based, according to Blomkvist and Hallin (2014), on two recurring activities: prototyping and presenting. The 4-phase model is based on two recurring activities: prototyping and presenting (Blomkvist & Hallin, 2014).	Blomkvist, P. & Hallin, A. (2014). *Method for Technology Students: Degree Projects Using the 4-phase Model*. Lund: Studentlitteratur.

Oxford system	In running text	In the reference list
	The 4-phase model is based, according to Blomkvist and Hallin, on two recurring activities: prototyping and presenting.[1]	Blomkvist, Pär & Hallin, Anette. *Method for Technology Students: Degree Projects Using the 4-phase Model*. Studentlitteratur, 2014.

[1] Pär Blomkvist & Anette Hallin. *Method for Technology Students: Degree Projects Using the 4-phase Model*. Lund: Studentlitteratur, 2014.

consultation with your supervisor (maybe there are special rules where you are). The only real requirement is that the list has a clear and logical structure that reflects the material you have actually used. If something has been deleted from the final version of the paper, then this material will also have to be deleted from the reference list.

There are many different ways of specifying references in a piece of academic work. The various subject fields all have their own favourites. Look at which system is being used in the scientific journals you are reading for your thesis and, if necessary, consult your supervisor about which referencing system best matches your work. The same simple rule applies here: choose a clear, easily-understood, and logical system for referencing and use it consistently throughout your entire paper.

The most common systems used in the social sciences are:

- The Harvard system, and the variant known as the APA system, and
- The Oxford system.

While you specify references in running text under the Harvard and APA systems, you place them in notes (most frequently footnotes, i.e. at the very bottom of the page, otherwise in endnotes) under the Oxford system. The two systems also differ as regards to how you arrange the references in the reference list (see illustration).

A very important thing to remember, and one which can save you lots of hours of work, is to specify references on a continuous basis and in all the documents you write. It is very easy to forget where you have retrieved a quotation or a piece of information if you do not keep tabs on its reference. In the worst case scenario, you might get it into your head that it is your own text and thus run the risk of being accused of plagiarism. Regardless of which referencing system you use in the finished paper, it may be a good idea to initially use the variant of the Oxford system available in Word. This is an automatic footnote system that links the reference to the text you have written. If you then move that text to another position within the document, the footnote will go with it (and be renumbered) and you will never need to worry about forgetting a source. There are also a number of different software programs for reference management which function similarly. Find out whether or not you, as a student, have access to one of these via the student's union or the IT department of your university.

> Choose a clear, easily-understood, and logical system for referencing and use it consistently.

ARGUMENTATION AND A THESIS DRIVEN PAPER

The finished academic has to pursue a clear thesis (the "thesis", in this meaning of the word, is your most important message to the reader, your proposition, formulated in a condensed way). This thesis, however, is not identical to the purpose of the investigation. You can, for instance, have as your purpose to investigate which future car fuel will become dominant in Europe. The thesis that you pursue will then be the fuel that you yourself advocate – i.e. your principal finding on the basis of the scientific investigation. The arguments presented in favour of the thesis must be well-founded (with evidence) in the empirical material investigated, and be backed up by both theory and previous research. It is not a matter of expressing opinions of your own regarding a topic. You will not know which thesis you are advocating until you have conducted your scientific study.

> You will not know which thesis you are advocating until you have conducted your scientific study.

> **Being surprised by the thesis**
>
> During a degree project, so-called silo effects were investigated within an organization. The students were pursuing the thesis that these problems could be remedied using cross-functional teams and improved technical systems for knowledge management. However, during the final phase of this work, they discovered that the vertical silo effects existing within the company ran the risk of being replaced by horizontal silo effects if their solution were to be implemented. Their thesis was changed radically by this insight and they were forced to take into account and rectify this newly-discovered set of problems. The company was surprised and very satisfied with their findings. This degree project received the highest grade.

Many students believe that when they advocate something they will be expressing their own opinions. But this is not the case in scientific contexts. Advocating something entails pursuing a thesis that can be shown to be well-grounded in the empirical study which has been conducted and which is in line with previous research and the theoretical specializations that exist within your own research field. It will absolutely not do to express your own feelings. If your argumentation and thesis go against *normal science*, i.e. scientifically-established thinking (paradigm), there will be even greater reason to present well-founded evidence of what you are asserting. One example of this is Charles Darwin, who hesitated for many years regarding whether or not he would dare to advocate his evolution theory since, in many ways, it went against the religious and scientific thinking of the time.

THE DEGREE PROJECT IN A WIDER CONTEXT – SCIENCE AND SOCIETY

Dimensions of power

The social sciences are interested in issues that concern people and the interplay between them. Dimensions of power are consequently of great significance in social science research and theory formation. Relationships of power are often analysed using terms like gender, class, ownership structure, globalization, and ethnicity. As a degree project author specializing in the social sciences, you should be aware that, by virtue of your choice of subject, you will end up in the midst of a social context in which these relationships of power affect both your practice and your interpretation of your findings. We recommend that you take into account the dimension of power when gathering your empirical material, when interpreting it, and when delivering your findings. One suitable way is to use the dimensions internal–external and justification–evaluation (see below) when critically analysing your own research process and product.

> It will absolutely not do to express your own feelings.

Our experience is that many students, before starting work on their degree projects, are entirely uninterested in, for instance, gender issues. "It doesn't concern me", "Society has changed – it's not like that anymore", are two examples of comments we often get. Unfortunately, that is not true, which many people also realise when starting working life.

Regardless of what you think, for example, of gender issues, you should take them into consideration during your work. If you are writing a degree project in which gender problems form a part of the purpose and the research question, this will, of course, be a given. But also for those not focusing on gender as a subject, the issue has a part to play – and the same applies to the dimensions of power that we mentioned above: class, ownership structure, globalization, and ethnicity.

Justification and evaluation

During Phase 3, you need to keep the criteria for science and *scientific quality* in mind the whole time. A *critical approach* has to be applied, as a general quality check, to the research you are conducting and to your product. It is appropriate to distinguish between four different dimensions when you are critically reviewing your own (and others') scientific work. It is partly a matter of how you can justify your study in an intrascientific way and externally, and partly a matter of how you should evaluate the findings in an intrascientific way and externally:

- *Intrascientific justification*. In this dimension of critical review, it is a matter of analysing whether or not the problematization and research focus can be justified from a scientific perspective. Does the investigation add anything new?
- *External justification*. Here, the question is posed of whether or not there is any societal value in your problematization. If you

have an external client, the answer is a given – the company/organization sees benefit in your investigation. But this does not suffice in terms of justification. You should also reflect upon how your work can contribute knowledge to society in general.
- *Intrascientific evaluation*. When your work is coming to an end, it will have to be evaluated using the scientific quality measures (reliability–validity, explanandum–explanans etc.).
- *External evaluation*. We think that you, as a researcher, should evaluate your findings from an external perspective as well. Once again, partly on the basis of the client's outlook – is the client satisfied? – and partly from an external and societal perspective – will your findings and their consequences benefit a large group of people?

The theory of science

It is important that you are aware of the fact that all theories have a number of points of departure on which the respective theory rests. These points of departure are in part assumptions about the type of phenomenon that the theory aims to understand or explain and in part assumptions about how knowledge of the phenomenon can, or should, be developed.

The first type of point of departure, which concerns assumptions about the phenomenon, can for instance relate to whether or not the phenomenon exists independently of the one observing it, or it if it comes about in connection with someone naming it or studying it. Another example is whether or not a phenomenon is static, or has been undergoing constant change. This type of point of departure is called the *ontological point of departure* regarding a phenomenon.

The second type of point of departure regarding the phenomenon that a theory aims to explain deals with how we can (should) study the phenomenon– how we develop new knowledge of it. Certain theories incorporate the outlook that new knowledge of the phenomenon primarily evolves from statistical studies in which a large volume of data is gathered regarding examples representing the phenomenon, while other theories pre-require an outlook whereby new knowledge primarily evolves by means of penetrating studies of individual examples: cases. The first outlook is often linked to the ideal that knowledge evolves in the best way when the researcher is objective in his/her relationship with that phenomenon, while the second is more

> **Positions in the theory of science**
>
> In the theory of knowledge and science, there are a great many conflicting answers to both the ontological and the epistemological question. Listed below are four extreme positions which are probably not entirely accepted or recommended by anyone:
>
> - *Realistic ontology.* Reality exists and looks like it does independently of whether we observe it or not.
> - *Constructivist ontology.* How we perceive reality is dependent on how we think about it and how we communicate with each other about it.
> - *Explanatory epistemology* (sometimes called positivistic). Independent knowledge acquisition is possible.
> - *Understanding epistemology* (sometimes called hermeneutic). Knowledge is dependent on the researcher's pre-understanding and interpretations.

strongly linked to the notion that knowledge evolves best when the researcher uses him-/herself as a tool during his/her research, but in a way that includes him/her being aware of his/her subjectivity. This type of question formulation has to do with *epistemological assumptions* regarding how we obtain our knowledge of reality.

The two types of assumptions described above deal with what is usually called *the theory of science*. This is a field that concerns issues applying to the nature of reality (ontology) and to how we can have some knowledge of this potential reality (epistemology). It is entirely possible that you will run into different theory of science specializations when working on your academic paper, perhaps primarily at seminars or method lectures. Some are also to be found here schematically described. However, we will not be getting involved in these in detail, instead we encourage you to seek knowledge of them elsewhere, depending on what is relevant to you in your work. It is good to be able to recognise different theory of science points of departure so that you may be able to determine whether or not it is possible to combine the different theories you run into and are thinking about using in your paper.

However, it is only in very special cases that you yourself will need to use theory of science terminology in order to increase the understanding of your research.

In reality, it will only be necessary when the purpose and question formulations focus on theory of science fields. If you are not writing that kind of paper, we feel that the terms should be left out. Your supervisor may show an interest in you positioning yourself from a theory of science point of view, and may thus ask you to do that in some, often early, prototype of the paper; however, in the completed thesis, this type of text is seldom required.

Explain or understand plus the issue of values

Regardless of whether or not you explicitly use theory of science terms, you should clarify if you, in your research, have the ambition to explain or understand, in addition to how your own values affect your scientific work.

There has been a long-lasting and infected debate in the social sciences regarding the terms *explain* and *understand*. One standpoint claims that explanations pre-require regulated causal connections (cause–effect) and that the social scientist can thus never, in a strict sense, have the ambition to explain. What remains is to understand what you are investigating. To some, the debate turns on having different definitions of the terms. You, as a degree project author, will in most cases not need to get involved in all the details regarding explanation or understanding. But it is important that you clearly decide which ambition you have and how you yourself define these disputed terms.

With regard to *values,* too, the debate is heated. Can the social sciences (or any sciences) provide objective answers? Is there a difference between natural science and social science regarding objectivity and the researcher's possibilities of being impartial vis-à-vis his/her subject? Once again, this involves definitions of the terms, and you will have to reflect upon your own standpoint. We recommend that you follow the instructions of the economist and Nobel laureate Gunnar Myrdal. He claimed that, even though it is not possible for the social sciences to be entirely devoid of value judgements, the researcher is still required to clearly account for the value premises guiding his/her investigation. If your values are capable of impacting upon

> " Even though it is not possible for social sciences to be entirely devoid of value judgements, the researcher is still required to clearly account for the value premises guiding his/her investigation.

your investigation, interpretation, and findings, then you must discuss this impact in your thesis.

Structure or actor?

While you are thinking about which theories and theoretical concepts you are using, the already-mentioned system perspective is of key importance. You must have theories that match your object of study, and when you are studying a phenomenon on the individual/organizational level, the functional level, or the industrial level. But to this we have to add another important distinction: that between structure focus and actor focus.

Having a structure focus or an actor focus in respect of your choice of theory is intimately linked with the following currently well-known questions:

- What is it that I want to understand/explain (explanandum)?
- What do I use in order to understand/explain (explanans)?

If you choose theories from the *structure focus*, you are interested in explaining or understanding phenomena which are connected, in some sense, with overarching structural circumstances. You can look for these regardless of which system level you are investigating. All of the following examples belong to the structural focus:

- You want to understand why certain individuals in an organization receive lower wages for doing similar work using the general gender theory of power relations between the sexes.
- You want to explain why the car industry chooses to implement Lean and you turn to theories of how ideas and technologies are disseminated between different industries or parts of the world
- You are interested in understanding why an entire industry changes and you use theories of industrial dynamic processes.

If instead you are primarily interested in how certain actors, i.e. individuals or organizations (these can also be seen as acting subjects), exert an influence on a course of events, then your research is *actor-focused*:

- You want to investigate why certain characteristics make you a good leader and you use theories taken from psychology or behavioural science.
- You want to understand which actors drive development when a new supply chain is about to be implemented at a company and you use theories from management and organization research.
- You want to understand why electric cars are having problems making a breakthrough and you choose theories of the influence of lobby groups on politics and industrial development.

It is not always easy to distinguish between the structure focus and the actor focus. Neither is it the case that they are mutually exclusive. It can work very well to combine the different focuses and discuss what is usually called *the interaction between structure and actor*, i.e. how societal structures affect the actor and how actors can affect and change structures.

However, it is important, on an analytical level and for the sake of clarity, to keep both foci apart – especially at the beginning of a research or degree project. It may be the case that you believe that theories and concepts should be taken from the structural focus, but then it turns out to be the case that the most important factors in explaining and understanding the phenomenon you are investigating are part of the actor focus. In any event, you have to be careful to keep these concepts apart and to carefully explain to the reader how you have reasoned.

> It can work very well to combine the different focuses and discuss what is usually called the interaction between structure and actor.

Phase 4 Deliver

Phase 4 is based on the insight that the engineer of today, and the future, not only has to have the capacity to produce credible and scientifically-proven decision-making data but also the ability to communicate these in various forms, both orally and in writing, to different types of recipients.

The Delivery Phase of the 4-phase model has also been inspired by something that we learnt as lecturers and supervisors: i.e. that the thesis often gets better if you link an oral presentation to your work on the written text. This entails that the prototype development during this phase is based on a number of iterations in which you return to the scientific tools and test your versions by presenting them to others (see Figure 16).

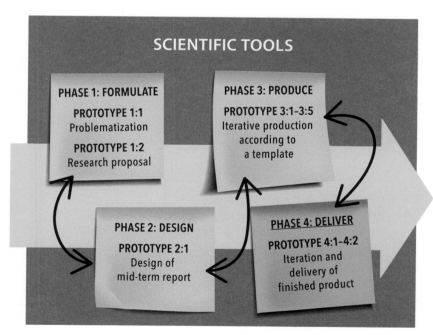

FIGURE 16 **Phase 4: Deliver**. In the last phase you must finalize your thesis, present it in a final seminar and deliver the written and oral presentation to your client and other stake holders. We have included two prototypes, 4:1 and 4:2, but there may be more. Also in phase 4 it's important to work iteratively; benefitting from written as well as oral presentations.

PHASE 4 DELIVER

Using the paper as a basis and a quality assurance tool, you will build other presentations, e.g. an executive summary, a PowerPoint presentation containing the income statement of the company, supportive data for workshops, proposals for implementing your findings, and further R&D based on your study (see Figure 16). But – and this is our point – you will be refining and specifying your arguments in the paper, too. Just as previous phases, the fourth phase will thus be about prototyping, feedback, and feedback-loops.

Prototype 4:1–4:2

The thesis, as already mentioned, will often be better if you link an oral presentation to your work on the text. Thus, we present below a number of steps sandwiching oral and written presentations (including Prototype 4:1–4:2). Naturally, the list is incomplete. It is possible to envisage many more situations where further prototypes, in addition to the finished thesis and its findings, are to be presented in different forms and to different audiences.

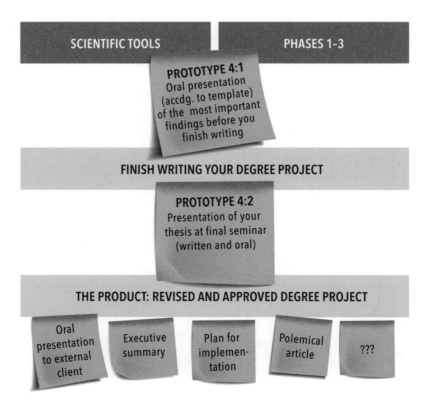

FIGURE 17 Prototype 4:1-4:2, with proposed workflow. Combine oral and written presentations. The results and your argument will become clearer and more powerful if you iaim at condensing the most important contributions of your thesis . Test different versions of your argument on different audiences and readers.

PROTOTYPE 4:1: ORAL PRESENTATION BEFORE THE THESIS IS FULLY FINISHED

In order to strengthen your arguments in the paper, it will be appropriate to deliver an oral presentation of the most important findings and your recommendations. This has to be done before you have written the final version of your degree project. We have noticed that academic papers get better when you add an oral presentation before you finish writing. Make sure you have a real audience, choosing from your client, fellow students, or relatives and friends. It does not matter too much who listens – the main thing is that someone does. Forcing yourself, over a limited period of time, to explain to an audience what you have actually arrived at hones your arguments and you will realise what is unimportant and can thus be cut out.

> **Kill your darlings**
>
> A major Swedish technology consultancy commissioned a degree project author to investigate the implementation of its sustainability strategy within the organization. The problem formulation focused on a survey of how the concept of sustainability was perceived by the employees and how it could become a part of the company's business model.
>
> The degree project author did the survey and gathered rich empirical material which, in and of itself, was enough to enable her to write her expected thesis. However, with just a month remaining, she realised that what she had been working on thus far was really just a pre-study for the really interesting problem that she had discovered: i.e. that different departments at the company had entirely different possibilities of delivering sustainable consultancy services.
>
> After a certain amount of anguish, she changed her purpose to investigating which strategies different types of technology consultants at the company were able to make use of in relation to the concept of sustainability. This also forced her to delete a lot of the text in her already-conducted study. Instead, the strategy opening became her main purpose and, in doing so, took pride of place in her thesis. It was a stressful final month but the degree project earned her a high grade and the student was directly offered a job at the company in order to continue developing the strategy that she had proposed.

PHASE 4 DELIVER

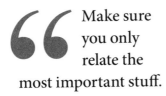

" Make sure you only relate the most important stuff.

Prepare a talk strictly in accordance with the template below (see pp. 146–147) – a classical template for an argumentative talk (which matches this speaking situation excellently as you have to convince your audience of the correctness of your findings). Give yourself a maximum of 20 minutes. Make sure you only relate the most important stuff, what needs to be said in order for your audience to understand what your investigation is about, what you have arrived at, and which recommendations you are delivering. Do not talk about everything in your degree project. Just focus on the findings because the talk exists in order for you to strengthen the arguments in your paper.

Ask your audience to take notes and to give you feedback regarding what they did not understand and things that were unclear, illogical, or completely missing.

Once you have delivered your talk, you will process the comments you receive from your audience together with your own insights from working on that talk. Remember that if your audience did not understand what you meant, then this will probably not depend on their not paying attention but most likely on your explaining things in a bad and unstructured way.

Now you have to write the final version of your degree project. Use the template you have chosen as a checklist. Be hard on yourself, *kill your darlings,* and prioritize continuity and logical consistency. Also use the learning goals as a checklist and make sure to achieve all the goals.

Return to the scientific tools and reconcile your work with the quality criteria specified. Be particularly careful with explanandum–explanans, as well as validity in relation to choice of theory, method, and sources. Are you investigating what you said you would, and are theory, method, and empirics working towards providing you with the answers you are seeking?

PROTOTYPE 4:2: PRESENTING AT A FINAL SEMINAR

An important part of the working method of academia is presenting degree projects at seminars. The writer is called the *respondent* while the reader of and commentator is called the *opponent* (you yourself will also be the opponent of someone else's thesis).

There are many rituals and templates regarding how a final seminar should be conducted, and regarding how the opponent and respondent are supposed to act. All universities and the various institutions have their own ways. In good time beforehand, investigate what is expected of you and discuss this matter with your supervisor.

A fundamental rule is that the seminar must be a respectful conversation between opponent and respondent. This is a matter of a *dialogue* between a serious critic, who takes his/her task seriously and who does not concentrate on finding minor errors but instead puts some effort into improving the thesis, and a respondent who is willing to listen and who is open to views. We think that the text you deliver to the final seminar must be one of a series of prototypes, even if the last, and one which is an almost finished product.

> A fundamental rule is that the seminar must be a respectful conversation between opponent and respondent.

Frequently, the respondent gets to deliver a brief oral presentation. This talk should not be arranged as strictly in accordance with the argumentative talk template as the one you delivered before the report was ready. In all probability, at the final seminar, you will be expected to dwell more on the introduction: problematization, purpose, question formulations, theory, and method. Your assignment as an opponent is to highlight the good points and to call attention to its shortcomings, while your assignment as a respondent is to take on board the criticism (see "Template for analysing academic texts", pp. 146–147).

Be open about both strengths and weaknesses. A thesis is graded in accordance with the quality of the final product, but is also based on the work process, the scientific depth, and the maturity of the author. You must therefore adhere to the academic code of honour regarding openness, transparency, critical awareness, and reflection.

Following the final seminar, you will most often be given time to revise and rectify some of the points arising during the discussion with your opponent and the seminar participants. Different seats of learning have different rules regarding the point in time at which the grade is awarded, but minor changes are usually allowed by the majority. Return once again to the learning goals and tick the ones you have achieved.

> You must therefore adhere to the academic code of honour regarding openness, transparency, critical awareness, and reflection.

PRESENTATIONS OF THE FINISHED PRODUCT

Frequently, your client will want you to deliver further products based on your investigation and thesis. An *executive summary* is a summary intended for those who do not have time to read the entire paper This should be about 15 pages long and should focus on the problematization, purpose, question formulations, findings, and recommendations; however, it should also contain a discussion of the theory and method in order to assure the reader of the investigation's reliability and validity. This minor report is, to put it briefly, an expanded abstract.

An oral presentation will often be given, for example, to a management team at the company that has not been directly involved in the process. It is important to bear in mind that this audience will probably mostly be interested in your concrete findings and recommendations – what these will entail for their company. Use the template for an argumentative talk (see below) and your executive summary and write an effective talk. Remember that you will probably not have too much time at your disposal.

SCIENTIFIC TOOLS

TEMPLATE FOR AN ARGUMENTATIVE TALK

A classical argumentative talk consists of three parts. Exactly as in Template B for the academic paper, there is an introduction, an argumentation section, and a conclusion. However, the purpose of the three parts is not quite the same, with the arrangement also being slightly different (stricter). Time-wise, the main focus of the talk must be the argumentation section. This means that the introduction can use about 20 percent of the time, the argumentation about 70 percent, and the conclusion about 10 percent (of the 15–20 minutes available to you in which to speak).

The basic structure of an argumentative talk is presenting a proposition (often called thesis) and the three main arguments/aspects of that proposition. To each main argument belong facts, quotations, statistics, analogies, and stories etc. which underpin your arguments.

Introduction

In the introduction to your talk, you will introduce yourself (or your group), the subject of the talk, a background to the problematization, your propositionand your three arguments. You will have to find the time for all of this during the initial minutes of your talk.

The initial purpose of the introduction is to establish something called *ethos*, i.e. that you are perceived by your audience to be someone they can trust and really want to listen to.

The second purpose of the introduction is to present your proposition/*thesis*, i.e. the most important message for your audience – what you want them to remember or what you want them to do. As regards your degree project, the proposition is constituted by your most important findings and which implications these will have for your client. It must be presented early on as you will then be showing respect for your audience by means of saying exactly what it is that you wish to convince them of. Lastly in the introduction, and in connection with the proposition, you present your three arguments summarized into one sentence, entailing that you are now moving over to the argumentation section of the talk.

Argumentation

> Argumentation is the most important part of the talk.

Argumentation is the most important part of the talk. You have built up your ethos in the minds of your audience and presented your proposition and arguments. Now it is a matter of delivering. Your argumentation must be based on scientifically-grounded evidence – on what you created in your degree project. During your argumentation, you will have to use logic – *logos* – and not appeal to your audience's feelings.

The talk must drive a thesis/leading idea and have *three main arguments*. These three arguments must clearly support the proposition by means of dealing with three different aspects.

It may be appropriate to specify a counter-argument against the proposition or one of the three supportive arguments (known as a *refutation*), in order for you later on be able to deliver your own counter-argument against this. The refutation should be added between Arguments 1 and 2 or between Arguments 2 and 3. You should not end the argumentation section with a refutation against yourself.

Conclusion

Summarize the thesis/proposition and the three main arguments in one sentence one more time. Conclude, if necessary, with some form of emotional argument – *pathos* – but only if you feel that your audience is with you.

It is important for the three arguments you choose not to be too similar. For example, you should not claim that the EU is demanding measures in Argument 1 and that Swedish authorities are doing the same in Argument 2. Instead, the argument should be that the authorities are demanding measures and the evidence of this is examples from both the EU and Sweden. A good metaphor for the argumentative talk is the three-legged stool. The seat is your proposition and it must be borne by three strong legs – the arguments. If two of the legs are too close together, the stool will become unstable and will then fall over. The same thing will happen, of course, if one of the legs is weak and cannot take the load.

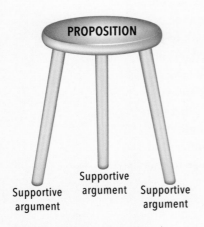

FIGURE 18 The three-legged stool.

The argumentative talk

Introduction
Presentation, background to problem etc. creating ethos:

> I want to advocate Volvo investing in plug-in hybrids. The reasons for this are; customers are demanding an environmental image, the price of oil is going to rise, and finally European authorities are demanding measures along those lines ...

Argumentation
Argument 1:

> Firstly, customers are demanding an environmental image ...

Evidence for Argument 1:

> In comparison with other car manufacturers and their environmental endeavours ... On the basis of statistics from other industries regarding green customer preferences (food) ...

Argument 2:

> Secondly, the price of oil is going to rise ...

Evidence for Argument 2:

> Statistics regarding the trend for the price of oil show that ...
>
> As is also emphasized in the debate regarding peak-oil ...

Refutation:

> It could be claimed that the debate regarding peak-oil is overstated
>
> But on the other hand ...

Argument 3:

> Thirdly, European authorities are demanding measures along those lines ...

Evidence for Argument 3:

> The most recent EU demands on vehicle manufacturers entail that ...
>
> Additionally, studies of other industries show that ...

Conclusion
Summarize the proposition and the main arguments and conclude by creating a touch of pathos:

> I have claimed that Volvo should invest in plug-in hybrids. The reasons for this are that its customers are demanding an environmental image, that the price of oil is going to rise, and, finally, that European authorities are demanding measures along those lines ... This investment will not just represent a good deal financially, we will also be doing our bit to prevent global warming.

Note that the example talk above takes up a subset of all the arguments that could support theproposition . We could, for instance, claim that Volvo's previous investments, or its special form of vehicle platform and vehicle portfolio, also provide good arguments for the proposition of investing in plug-in hybrids. Which arguments you choose will firstly depend on what kind of investigation you have conducted – not all investigations provide the same supportive data for argumentation. Secondly, it will depend on what kind of audience you are speaking to. The example talk can be said to be directed at an audience on a relatively high level within the company. If you are speaking to a more technically-aware audience (e.g. line managers, automotive engineers), you will then have to adapt your arguments to their knowledge and expectations, and to that situation. The most important piece of advice, besides preparing yourself and having a clear structure, is to adapt your talk to your audience.

TEMPLATE FOR ANALYSING ACADEMIC TEXTS

This template is intended to work as an aid in the reading and analysis of academic texts. It can be used when you are reading up on the scientific literature for your degree project as well as when you are acting as an opponent or critic of the prototypes of your fellow degree project students. The template functions as general supportive data for your reading and comments. The different stages of the framework are on different analysis levels. Stage 1 (purpose/question formulation) and Stage 2 (findings) are close to the text's concrete content and the author's intentions. In Stage 3 (evidence/argumentation), you assess whether or not the author is credible when putting forward his/her findings. In Stage 4 (proposition), your analysis is on another level than the actual findings and you make your *own* overarching interpretation of the message in the text. In Stage 5 (readability/relevance), you account for your personal impressions of the text.

Stage 1. Purpose/question formulation

Emphasize the author's own purpose in writing the text and the questions that he/she claims to want to answer. You are to account for the assignment that the author has given him-/herself. What does the author want to show? Which questions are to be answered? Is there a clear client/reader? Note that the text does not always clearly account for the author's question formulations.

Stage 2. Findings

Account for the most important empirical findings and for the conclusions that the author draws. Do the findings tally with the purpose? Do the questions posed receive any answers?

Stage 3. Evidence/argumentation

Which evidence is the author using? Is this evidence credible? How does the author advocate his/her findings? Are the arguments sustainable and credible?

Stage 4. Proposition

In your opinion, which proposition is the author's most important one? Is the proposition accounted for clearly? Here, it is a matter of fishing out one or more of the overarching messages that the author is conveying (consciously or unconsciously). Note that the proposition is not the same thing as the author's purpose. Account for the proposition briefly. Do you agree regarding the proposition?

Stage 5. Readability/relevance

Account for what you think of the text in terms of being a reading experience. Was the text interesting, easy, difficult? Is the text, in your opinion, relevant to the degree project?

CONCLUSION

Now that you have come this far, it will soon be time to hand in the final version of the text. You have probably taken on board the final opposition conducted by one or more of your coursemates and you have worked on your text in accordance with both their and your supervisor's comments. Different seats of learning have different rules regarding how many of the comments from the final seminar can be included in final manuscripts.

Some universities print degree projects and then you will have a process ahead of you during which you send your manuscript to the printers and get a galley proof in return where you make minor changes prior to publication. Frequently, your degree project will be assigned an ISRN number, i.e. a unique number which identifies your thesis. Some universities have other kinds of numbering systems. How things look at your university is something you will need to find out before sending your work to the printers.

Once you have worked your way through the 4 phases of your degree project, you will have completed a journey which at times was frustrating, but was hopefully also very educational. You will have learnt to formulate a problem on the basis of an often complex situational description, to search and summarize relevant literature in the field, to choose methods of data gathering, to gather and analyse data, and to draw conclusions. Regardless of whether or not you are interested in continuing to the next educational level – becoming a PhD student – you will benefit greatly from these lessons.

We believe that, during the course of working on your degree project, you yourself will have discovered, as we wrote in the introduction, the fact that many modern engineers are actually a kind of social engineer working in a multidisciplinary way in a field where science and technology meet society. We also believe that you will have discovered that both the prototyping and the work of conducting an investigation, not to mention writing up your findings, constitute a work of craftsmanship and that this is something that you learn by doing. When you hand in your thesis to the printers, you will have learnt the craftsmanship, something which you will benefit greatly from in the future.

Many engineers collaborate with researchers while even more follow the process described by the 4-phase model when conducting investigations or pre-studies. It is a certainty that you will continue investigating and writing reports. You will be forced to find and formulate problems, choose your investigative methods, and advocate both orally and in writing your ideas and choices in front of executive teams and boards in order for them to invest time and resources specifically in your project.

> Many engineers collaborate with researchers while even more follow the process described by the 4-phase model when conducting investigations or pre-studies.

Maybe it is not always a requirement for you to explicitly account for your method, theory, and all the demands that academia makes of reports. But in any case, your arguments must be well-grounded in both existing knowledge and your empirical investigation. The scientific demands are thus always present and constitute the quality assurance of your work. Consequently, we want to tone down the perception that universities want abstract theory while companies want applicable findings. We realise that this disparity can exist but we assert that it is significantly smaller than it may seem; above all, it is a superficial phenomenon that has to do with style rather than substance. Both academia and the corporate world require findings of high quality which they can trust (reliability and validity), even if the genres you are writing in can differ. All of your credibility as a researcher or consultant is thus contained within the scientific method of working. The academic paper is the mother of all documents and, in a very concrete way, your professional indemnity insurance should your findings be called into question.

We also hope that you, as a degree project author, have learnt to appreciate the iterative nature of research and that you have realised that you, as a researcher (or consultant), have to be open to the possibility of your purpose, research questions, empirics, interpretations, and so on changing several times during the course of prototyping while you are struggling to write increasingly precise formulations. Your purpose, etc., should not be set in stone too early in the process; this will almost certainly lead to a suboptimal result. Your investigation may easily become trivial if you do not allow yourself to be surprised by what you discover during the course of your journey.

Finally, we believe that you have discovered that a certain resilience to stress is required in order to be able to preserve multiplicity during the research process. But, as in all creative work, your satisfaction will be so much the greater when you discover linkages and findings that you previously had no idea about. It is both very enjoyable and a privilege to have the opportunity to make a real discovery and to contribute something new.

FURTHER READING

Prototyping

Blomkvist, P. & Uppvall, L. (2012). A chain is only as strong as its weakest link: Managing change in the curriculum of industrial management education. *International Journal of Industrial Engineering and Management* (IJIEM), 3 (2), pp. 53–65.

Blomkvist, P. & Uppvall, L. (2012). Learning to love ambiguity: Authentic live case methodology in industrial management education. *International Journal of Case Method Research & Application*, XXIV (4), pp. 272–285.

Carleton, T., Cockayne, W. & Tavahainen, A. (2013). *Playbook for strategic foresight & innovation.* Available at: http://foresight.stanford.edu:16080/playbook/.

Laws and rules applicable to degree projects

You can find more information about the laws and rules applicable to university degree projects at the website of the Office of the Chancellor of the Swedish Universities

Overview of method

Bryman, A, Bell, E (2011) *Business research methods* (3. ed.) Oxford : Oxford University Press

Overview of qualitative method

Berg, B.L. (2007). *Qualitative research methods for the social sciences.* (6th ed.) Boston: Pearson.

Prasad, P. (2005). *Crafting qualitative research: Working in the postpositivist traditions.* Armonk, NY & London: M.E. Sharpe.

FURTHER READING

Case study methodology

Eisenhardt, K.M. (1999). Building theories from case study research. *Academy of Management Review*, 14 (4), pp. 532–550.

Eisenhardt, K.M. & Graebner, M.E. (2007). Theory building from cases: Opportunities and challenges. *Academy of Management Journal*, 50 (1), pp. 25–32.

Flyvbjerg, B. (2006). Five misunderstandings about case-study research. *Qualitative Inquiry*, 12 (2), pp. 219–245.

Yin, Robert K. (2009) *Case study research : design and methods*. London: SAGE

Action research

Coghlan, D. & Brannick, T. (2010). *Doing action research in your own organization*. (3rd ed.) London: Sage.

Reason, P. & Bradbury, H. (ed.) (2006). *Handbook of action research*. Bradbury: Sage.

Observation methodology

Clifford, J. (1986). Introduction: Partial truths. I J. Clifford & G.E. Marcus (ed.), *Writing culture: The poetics and politics of ethnography*. Berkeley, Los Angeles, London: University of California Press, pp. 1–26.

Czarniawska, B. (2007). *Shadowing and other techniques for doing fieldwork in modern societies*. Malmö: Liber.

Golden-Biddle, K. & Locke, L. (1993). Appealing work: An investigation of how ethnographic texts convince. *Organization Science*, 4 (4), pp. 595–616.

Hobbs, D. & Wright, R. (ed.) (2006). *The Sage handbook of fieldwork*. London, Thousand Oaks, New Delhi: Sage.

Kozinets, R.V. (2010). *Netnography: Doing ethnographic research online*. Thousand Oaks, CA: Sage.

Kunda, G. (1992). *Engineering culture: Control and commitment in a high-tech corporation*. Philadelphia: Temple University Press.

Van Maanen, J. (1979). The fact of fiction in organizational ethnography. *Administrative Science Quarterly*, 24 (4), pp. 539–550.

Interview methodology

Kvale, S & Brinkmann, S (2009) *InterViews : learning the craft of qualitative research interviewing* (2. ed.) Los Angeles: Sage Publications

Survey methodology and quantitative analysis

Dillman, D.A., Smyth, J.D. & Christian, L.M. (2008). *Internet, mail, and mixed-mode surveys: The tailored design method*. Hoboken, NJ: Wiley.

Groves, R.M. et al. (2004). *Survey methodology.* Hoboken, NJ: Wiley.
Hair, J.F. Jr et al. (2013). *Multivariate data analysis: A global perspective.* (7th global ed.) Boston: Pearson.
Wooldridge, J. (2013). *Introductory econometrics.* Mason, OH: Cengage Learning.

Narrative method

Czarniawska, B. (2004). *Narratives in social science research.* London: Sage.

Discourse analysis

Jørgensen, M. & Phillips, L. (2002). *Discourse analysis as theory and method.* London: Sage.

Writing

Styhre, A. (2013). *How to write academic texts: A practical guide.* Lund: Studentlitteratur.

Referencing systems

The standard work for almost all referencing systems and other scientific writing rules is *The Chicago Manual of Style* (University of Chicago): http://www.chicagomanualofstyle.org/home.html.

Presenting and being an opponent

Björklund, M. & Paulsson, U. (2003). *Seminarieboken: Att skriva, presentera och opponera.* Lund: Studentlitteratur.
Hellspong, L. (2011). *Konsten att tala: Handbok i praktisk retorik.* (3rd ed.) Lund: Studentlitteratur.
Hägg, G. (2011). *Retorik i tiden: 23 historiska recept för framgång.* Stockholm: Norstedts.
Renberg, B. (2013). *Bra skrivet, väl talat: Handledning i skrivande och praktisk retorik.* (2nd revised ed.) Lund: Studentlitteratur.

Scientific legitimacy and theory

Alvesson, M. & Deetz, S. (2000). *Kritisk samhällsvetenskaplig metod.* Trans. S.-E. Torhell. Lund: Studentlitteratur.
Alvesson, M. & Sköldberg, K. (2008). *Tolkning och reflektion: Vetenskapsfilosofi och kvalitativ metod.* (2nd ed.) Lund: Studentlitteratur.

FURTHER READING

Bjereld, U., Demker, M. & Hinnfors, J. (2002). *Varför vetenskap? Om vikten av problem och teori i forskningsprocessen.* (2nd ed.) Lund: Studentlitteratur.

Gustavsson, B. (2002). *Vad är kunskap? En diskussion om praktisk och teoretisk kunskap.* Stockholm: National Agency for Education.

Hansson, B. (2011). *Skapa vetande: Vetenskapsteori från grunden.* Lund: Studentlitteratur.

Harnow Klausen, S. (2006). *Vad är vetenskap?* Trans. M. Hasselström. Stockholm: Natur & Kultur.

Myrdal, G. (1968). *Objektivitetsproblemet i samhällsforskningen.* Trans. B. Lindensjö & H. Silling. Stockholm: Rabén & Sjögren.

SUBJECT INDEX

4-phase model 9

abduction 48
abstract 104, 108
access 65, 68, 84
action research 72, 73
actor focus 135
alienate 84
alignment *See* logical consistency
analysis of empirics 114
analytical contribution 40
analytical generalizability 67
argumentation 144, 146
arithmetic reliability 53

being an opponent 53, 140, 141
best practice 40
bivariate analysis 120

case study 63, 65–68
causality 47
client 8, 10, 11, 33–35
cluster sample 70
confidentiality requirement 37
connections 47, 50, 69, 116, 134
consent requirement 37
consistency 96, 102
critical attitude 45, 54, 65
critical discourse analysis 118
critical thinking 54

data gathering 68, 74
deduction 48
degree project 33
demarcation 41

design 12–14, 57
dialogical reliability 53
discourse analysis 115, 118, 119, 124
disposition
 – template A 98, 100
 – template B 98, 111
document gathering 74
dropout analysis 94, 95

empirical contribution 40, 41
empirics 25, 48
epistemology 133
ethics 37, 38
ethos 143
evaluation 131
Excel 94, 119, 121, 122
executive summary 35, 138, 142
experiment 72, 125
explanandum 62, 63
explanans 62, 63
extrapolation 67

factor analysis 122
feedback 15
feedback-loop 15
focus group methodology 74
frequency 72, 94–96, 120

generalizability 66, 67, 71, 72
genre 48
going native 86

Harvard system 128
hypothesis 48

impartiality 35, 53
independence 37, 86
induction 48, 102
information requirement 37
ingress 103
internal reliability 96
interval variable 89, 120
interview methodology 74–76

justification 131, 132

kill your darlings 140
knowledge interest 36

language 48–51
layout 93, 107, 108
learning goals 39
literature 43–46
logical consistency 102, 103, 105, 109, 114, 140
logos 144

managerial implications 101
MECE 92
median 72, 120
method 25, 61–63
mid-term report 14, 57
multiplicity 75, 79
multivariate analysis 120

narrative analysis 115, 117, 118
netnography 82
nominal variable 120
non-trivial problem 21, 29

SUBJECT INDEX

objectivity 35, 79, 87, 134
ontology 133
operationalization 89, 90, 96
oral presentation 141–143
ordinal variable 120
outlier 119
Oxford system 128, 129

pathos 144, 145
peer review 24, 46, 53, 102
phenomenon-driven work 46
plagiarism 38, 46, 127, 129
population 70–72, 119
positioning 42, 44, 134
presenting 9
primary source 123, 127
problematization 20–23, 29
process 8
produce 97
prototyping 9
purpose 23, 26–29

qualitative/quantitative
   method 61
quantitative analysis 119
quantitative study 68
question formulation 26
quotes 38

ratio 89, 120
raw data 114
reference 108, 122
reference group 53
reflective writing 51, 114, 115,
   117–119
refutation 144
regression analysis 122
reliability 53, 96, 124
reproducibility 96
researchable problem 11, 19, 21
research design 26, 62
research diary 16
research proposal 31, 41
respondent 53, 140
response frequency 94–96

sampling 70, 71
saturation 77, 78
science 52
scientific legitimacy 25, 78
scientific quality 52, 131, 132
scientific tools 12
scientific writing 9
secondary source 123
snowball sampling 70
social engineer 7
source 38, 122–127

source criticism 122, 124
SPSS 94, 119, 121, 122
stability 96
standard deviation 71, 120
statistical generalizability 67, 71
structural focus 135, 136
style 48–50
summary 35, 45, 50, 104, 108,
   138, 142
survey methodology 74, 88
synopsis 103, 104
systematic sampling 70
system perspective 29–31, 135

telephone survey 95
thematic analysis 115, 116, 124
theoretical contribution 40
theory of science 132–134
thesis proposal 39

univariate analysis 120

validity 52, 53, 96, 126
variable 89, 120
variance 120
writing 48

über reading up 43